The City Besieged

Culture and History of the Ancient Near East

VOLUME 36

The City Besieged

Siege and Its Manifestations in the Ancient Near East

by

Israel Eph'al

BRILL

MAGNES

LEIDEN • BOSTON
2009

This book is printed on acid-free paper.

Library of Congress Cataloging-in-Publication Data

Eph'al, Israel.
[Ke'-ir netsurah. English.]
The city besieged : siege and its manifestations in the ancient Near East / by Israel Eph'al.
p. cm. — (Culture and history of the ancient Near East)
Includes bibliographical references and index.
1. Siege warfare—Middle East. 2. Sieges—Middle East. 3. Military art and science—Middle East—History. I. Title. II. Series.

UG444.E6413 2009
355.4'409394—dc22

2008048072

ISSN 1566-2055
ISBN 978 90 04 17410 8

PREFACE TO THE HEBREW EDITION

Many of my friends have asked me, and I myself have at times wondered, what led me to write a book about siege in warfare in the ancient Near East. To a considerable extent, I was motivated by a recognition (that has been with me for a long time, since my youth, throughout my military service and many years in the reserves) that the State of Israel, a small territory surrounded by enemies, is in effect under constant siege. I chose to spend my sabbatical leaves in North America rather than in Europe, in order to enjoy the vast, almost unlimited vistas of that continent and thereby somewhat relieve the sense of being constantly embattled and confined. The signings of peace treaties with Egypt and Jordan awakened the hope that the isolation of the State of Israel would gradually lessen, and that the feeling of siege that has accompanied us for over fifty years would disappear.

One of the restrictions I imposed upon myself in writing this book was to deal only with information based on ancient sources, and to refrain from discussions of undocumented theoretical aspects relating to the long history of siege in warfare. While these might deepen and enrich the sense of the experience of siege, they are not attested to in any of the ancient Near Eastern sources. Everything discussed here is based on definitive ancient finds of the period, up to the time of Alexander the Great, and is illuminated by military thinking, which is valid for all the different periods.

Mr. Elnathan Weissert was of great assistance to me in checking the references and quotations from Akkadian literature; his useful comments have left their impression throughout the work.

The Hebrew University of Jerusalem, 1996

PREFACE TO THE ENGLISH EDITION

This is an updated and expanded edition of a book that originally appeared in Hebrew in 1996. The expansion is reflected primarily in the addition of new sources and bibliographical items that have appeared since the Hebrew publication. Citations of various sources have been updated in the light of new editions published in recent years.

Eleven years have passed since the appearance of the Hebrew edition. Unfortunately, in the course of these years there has been no progress in terms of the isolated situation of the State of Israel. The hopes expressed in the preface to the Hebrew edition, that isolation would gradually diminish and with it the sense of siege that has beset Israel's inhabitants since its establishment, are as yet unfulfilled.

The Perry Foundation undertook to publish the present English edition of the book. The Society of Biblical Literature contributed partial funding of the translation. Dr. Dana Cohen has edited the English style. My thanks to them all.

Contents

ABBREVIATIONS

A	siglum of tablets in the collections of the Oriental Institute, University of Chicago, or of texts from the Mari excavations at the Musée du Louvre, Paris
ABL	Harper, R. F., *Assyrian and Babylonian Letters Belonging to the K(ouyunjik), Collection(s) of the British Museum*
ACh	Virolleaud, Ch., *L'astrologie chaldéene*
AfO	*Archiv für Orientforschung*
AHw	W. von Soden, *Akkadisches Handwörterbuch*
AJBI	*Annual of the Japanese Biblical Institute*
AJSL	*American Journal of Semitic Languages and Literatures*
ANEP	Pritchard, J. B., *The Ancient Near East in Pictures*
ANET	Pritchard, J. B. (ed.), *Ancient Near Eastern Texts Relating to the Old Testament*
AnOr 8	Pohl, A., *Neubabylonische Rechtsurkunden aus den Berliner Staatliche Museen*, I
AnSt	*Anatolian Studies*
AOF	*Altorientalische Forschungen*
ARM I	Dossin, G., *Correspondance de Šamši-Addu et ses fils*
ARM II	Jean, Ch.-F., *Lettres diverses*
ARM XXVI	Charpin, D., Joannès, F., Lackenbacher, S. and Lafont, B., *Archives épistolaires de Mari* I/1-2
ARM XXVII	Birot, M., *Correpondance des gouverneurs de Qaṭṭunân*
ASJ	*Acta Sumerologica*
AuOr	*Aula Orientalis*
BaM	*Baghdader Mitteilungen*
BASOR	*Bulletin of the American Schools of Oriental Research*
BiOr	*Bibliotheca Orientalis*
BJPES	*Bulletin of the Jewish Palestine Exploration Society* (Hebrew)
BM	Siglum of objects in the British Museum, London

Borger, *Asarh.*	Borger, R., *Die Inschriften Asarhaddons, Königs von Assyrien*
BR 8/7	San Nicolò, M., *Babylonische Rechtsurkunden des ausgehenden 8. und des 7. Jahrhunderts v. Chr.*
BSA	*Annual of the British School of Archaeology at Athens*
CAD	*The Assyrian Dictionary of the University of Chicago*
CAT	Dietrich, M., Loretz, O. and Sanmartin, J., *The Cuneiform Alphabetic Texts from Ugarit, Ras Ibn Hani and Other Places*
CT 30	Handcock, P. S. P., *Cuneiform Texts from the Babylonian Tablets in the British Museum*
CTN II	Postgate, J. N., *The Governor's Palace Archive*
EA	Knudtzon, J. A., *Die El-Amarna-Tafeln* (English edition: Moran, W. L., *The Amarna Letters*)
EI	*Eretz-Israel*
Enc. Miqr.	*Encyclopedia Miqra'it (Encyclopaedia Biblica)*
ET	Goodnick Westenholz, J., *Cuneiform Inscriptions in the Collection of the Bible Lands Museum, Jerusalem: The Emar Tablets*
GCCI II	Dougherty, R. Ph., *Archives from Erech, Time of Nebuchadrezzar and Nabonidus*
Grayson, *Chron.*	Grayson, A. K., *Assyrian and Babylonian Chronicles*
IEJ	*Israel Exploration Journal*
IM	siglum of tablets in the Iraq Museum, Baghdad
IOS	*Israel Oriental Studies*
ITP	Tadmor, H., *The Inscriptions of Tiglath-pileser III King of Assyria*
Izbu	Leichty, E., *The Omen Series šumma izbu*
JANES	*Journal of the Ancient Near Eastern Society of Columbia University*
JARCE	*Journal of the American Research Center in Egypt*
JCS	*Journal of Cuneiform Studies*
JESHO	*Journal of the Economic and Social History of the Orient*
JNSL	*Journal of Northwest Semitic Languages*
JQR	*Jewish Quarterly Review*
JSOT	*Journal for the Study of the Old Testament*
JSSEA	*Journal of the Society for the Study of Egyptian Antiquities*

JTVI	*Journal of the Transactions of the Victoria Institute*
K	siglum of tablets in the Kouyounjik Collection at the British Museum
KAI	Donner, H. and Röllig, W., *Kanaanäische und armäische Inschriften*
KAR	Ebeling, E., *Keilschrifttexte aus Assur religiösen Inhalts*
KBo	Keilschrifttexte aus Boghazköi
Lie, *Sar.*	Lie, A. G., *The Inscriptions of Sargon II, King of Assyria*, Part I: *The Annals*
MARI	*Mari: annales de researches interdisciplinaire*
MIO	*Mitteilungen des Instituts für Orientforschung*
NABU	*Nouvelles assyriologiques brèves et utilitaires*
NALD	Postgate, J. N., *Fifty Neo-Assyrian Legal Documents*
NATAPA I	Fales, F. M. and Jakob-Rost, L., 'Neo-Assyrian Texts from Assur: Private Archives in the Vorderasiatisches Museum of Berlin, Part I,' *SAAB* 5 (1991), 3-157
NCBT	siglum of the Newell Collection of Babylonian Tablets, Yale University, New Haven
ND	siglim of tablets from the British excavations at Nimrud (Calah)
NL	Nimrud Letter
OECT 10	McEwan, G. J. P., *Late Babylonian Texts in the Ashmolean Museum*
OIP II	Luckenbill, D. D., *The Annals of Sennacherib*
OLP	*Orientalia Lovaniensia Periodica*
OTS	*Oudtestamentische Studien*
PEF QSt	*Palestine Exploration Fund Quarterly Statement*
RA	*Revue d'assyriologie et d'archéologie orientale*
RAI	*Compte rendu, Rencontre Assyriologique Internationale*
RB	*Revue biblique*
RDAC	*Report of the Department of Antiquities, Cyprus*
RE	Beckman, G., *Texts from the Vicinity of Emar in the Collection of Jonathan Rosen*
REA	*Revue des études anciennes*
RIDA	*Revue internationale des droit de l'antiquité*
RIMA 1	Grayson, A. K., *Assyrian Rulers of the Third and Second Millennia BC (to 1115 BC)*

RIMA 2	Grayson, A. K., *Assyrian Rulers of the Early First Millennium BC, I (1114-859 BC)*
RIMA 3	Grayson, A. K., *Assyrian Rulers of the Early First Millennium BC, II (858-745 BC)*
RPAE	Arnaud, D., *Recherches au pays d'Aštata. Emar* VI/1-4
RSF	*Rivista di studi fenici*
RT	*Recueil de travaux relatifs à la philologie et à l'archéologie égyptiennes et assyriennes*
SAA II	Parpola, S. and Watanabe, K., *Neo-Assyrian Treaties and Loyalty-oaths*
SAA III	Livingstone, A., *Court Poetry and Literary Miscellanea*
SAA IV	Starr, I., *Queries to the Sungod: Divination and Politics in Sargonid Assyria*, Helsinki 1990.
SAA VI	Kwasman, Th. and Parpola, S., *Legal Transactions of the Royal Court of Nineveh,* Part I: *Tiglath-Pileser III through Esarhaddon*, Helsinki 1991
SAA X	Fuchs, A. and Parpola, S., *The Correspondence of Sargon II,* Part III
SAAB	*State Archives of Assyria Bulletin*
Shemshara	Eidem, J. and Laessøe, J., *The Shemshara Archive*, 1. *The Letters*
Shnaton	*An Annual for Biblical and Ancient Near Eastern Studies* (Hebrew)
SMEA	*Studi Micenei ed Egeo-Anatolici*
Streck, *Asb.*	Streck, M., *Assurbanipal und die letzten assyrischen Könige bis zum Untergange Niniveh's*
TA	*Tel Aviv*
TCL III	Thureau-Dangin, *Une relation de la huitième campagne de Sargon*
TCL VI	Thureau-Dangin, F., *Tablettes d'Uruk à l'usage des prêtres du temple d'Anu au temps des Sèleucides*
TSABR	Arnaud, D., *Textes syriens de l'âge du bronze récent*
TUAT	Kaiser, O. et al. (eds.), *Texte aus der Umwelt des Alten Testaments*
UCP	*University of California Publications*
UF	*Ugarit-Forschungen*
VT	*Vetus Testamentum*
WO	*Die Welt des Orients*

WZKM	*Wiener Zeitschrift für die Kunde des Morgenlandes*
YOS VI	Dougherty, R. Ph., *Records from Erech, Time of Nabonidus (555-538 B.C.)*
YOS X	Goetze, A., *Old Babylonian Omen Texts*
ZA	*Zeitschrift für Assyriologie*
ZAW	*Zeitschrift für die alttestamentliche Wissenschaft*

CHAPTER ONE

INTRODUCTION

I. WHAT IS SIEGE WARFARE?

Siege is a form of warfare in which one of the combative sides defends itself within an area delimited by a system of obstacles, while the opponent attempts to penetrate these obstacles and to engage in hand-to-hand combat, in which its superiority is assured. Unlike pitched battle, which is generally dynamic and brief and in which mobility plays a significant role, siege warfare is protracted and static by nature.

The decision regarding the form of warfare, that is, the determination that a given confrontation between the two parties will take the form of siege rather than that of pitched battle, lies in the hands of the defending party. Generally, such a decision is taken when the defender estimates that his chances of success on the battlefield are slim, and therefore makes an effort to refrain from direct engagement. The attacker determines the technical nature of the siege – whether a breakthrough to the protected area is attempted or a blockade is imposed, which entails the risk of starvation, thirst, and disease for the besieged. This decision depends on the resources at the attacker's disposal, including time, technical means, the cost in manpower and the availability of breakthrough materiel.[1]

1 Unlike the breakthrough, the blockade entails loss of time and involves supply problems for the attackers, but minimizes the loss of manpower; Cf. Plutarch, *Pericles*, 27: "He (Pericles) conquered and routed the enemy and at once walled in their city (Samos), preferring to get the upper hand and capture it at a price of money and time, rather than of the wounds and deadly perils of his fellow-citizens (the Athenians)." By contrast, armies of far-flung empires, with several potential areas of conflict, could not afford extended siege battles – due to factors of time and space – and were forced to break

As noted above, in siege warfare the attacker needs to overcome obstacles, whether natural (such as a river or a natural slope) or artificial (including passive obstacles, such as moats, walls, barbed wire or minefields, or active ones, such as projectiles of various weapons, from the sling and the bow to modern-day automatic firearms and artillery). Surprise tactics and deception or exploitation of inside intelligence facilitate rapid and easy penetration into the protected area. Any other method to attain this objective involves breakthrough - forceful penetration, frequently requiring an extended effort. Over the ages, technological development naturally led to the emergence of suitable techniques of breakthrough; yet, earlier obstacles (e.g., moats and walls) are similar in principle to modern ones (e.g., minefields or fire zones), with the exception of modern-day vertical attack by airborne combatants. At every turn, the attacker takes advantage of his superiority in place and time to effect the breakthrough. Therefore, where there is a continuous series of obstacles, penetration would be attempted at a small section of the defense line, in which the attacker can concentrate efforts and make a successful advance despite opposition.

Once the defender has discovered the breaching point, he will attempt to block it, generally leaving little time for the attacker to penetrate the breach. The breaching stage is, therefore, decisive for both attacker and defender. Consequently, both sides will attempt to mobilize their best commanders and fighters in the break-in zone at this stage.

Unlike modern-day warfare, in which engines, radio and explosives are extensively used, ancient siege warfare involved a slow process of overcoming obstacles; therefore, it was not always possible to concentrate on a single method of breakthrough. On more than one occasion, several different methods and a variety of assault equipment were simultaneously employed in the hope that one of them would succeed.

Siege warfare is a universal phenomenon, many of whose manifestations are connected to the human spirit, thus revealing

---

into the besieged cities. SeeEphʿal in *History, Historiography and Interpretation*, 91-104.

similarities across place and time. What, then, is unique about siege warfare in the ancient Near East (up to the arrival of Alexander the Great in the region, in 332 BC)? The answer lies in its military technology, in the specific socio-economic reality, in the behavior of the general population and of the individual within a unique framework of custom and ideology and in the range of sources and genres from which we derive our information about siege warfare during that period.

II. THE PURPOSE OF THIS BOOK AND METHOD OF DISCUSSION

To date, there has only been limited discussion of the phenomenon of siege warfare in the ancient Near East. Comprehensive studies that go beyond specific siege battles are included in the works of Waschow,[2] Yadin,[3] and Oppenheim.[4] The first two are devoted to the military techniques characteristic of this type of warfare, while the third deals with legal formulae found in the Babylonian "siege documents." The evidence (particularly written evidence) included in the works of Waschow and Yadin is sparse and sporadic, and many topics are not addressed at all. Much information has accumulated since the appearance of these pioneering studies, information through which we can extend our knowledge of the topics already examined and explore new ones.

The present book differs from its predecessors primarily in the underlying approach. It is contended that the phenomenon of siege warfare is extremely complex and cannot be properly perceived or understood without familiarity with the range of facets involved and an awareness that these facets were all simultaneously present. Accordingly, the picture presented in this book has been expanded to include various

2 Waschow, *4000 Jahre Kampf um die Mauer: Der Festungskrieg der Pioniere*.

3 Yadin, *The Art of Warfare in Biblical Lands in the Light of Archaeological Study*.

4 Oppenheim, "'Siege Documents' from Nippur," *Iraq* 17 (1955), 69–89.

aspects relating to siege warfare: military, economic, legal, social and literary. This approach is vital to the understanding of the phenomenon in concrete and abstract terms alike.

The discussion is delimited to siege warfare in the ancient Near East – the period prior to the victory of Alexander the Great over King Darius III of Persia, which marks the beginning of the Hellenistic period in the Near East (331 BC). This decision is based on convenience rather than on rigorous methodological considerations: after all, data pertaining to the subjects of this book can be found in sources outside the ancient Near East, from centuries or even millennia later than the period under discussion, and includes various topics not discussed here in the absence of suitable documentation in ancient sources. Chronology is of significance to our discussion only in subjects related to technology, be it military or civilian. With respect to the principles of war, the human spirit and the behavior of individuals and groups, chronology is of no particular significance. Our knowledge of later and better documented sieges, such as those of La Rochelle (1627–1628) or of Leningrad (1941),[5] may thus be of help in reconstructing the social and economic reality of siege in the ancient Near East.

The extant written sources concerning siege are primarily literary, and generally provide only the most basic information concerning the actual technique of siege warfare. More detailed and varied information concerning the technical aspects of this form of warfare emerges from archaeological finds and representations in the plastic arts. Examination of these data – including changes in the structure of fortifications (e.g., the use of offset-and-inset walls on the one hand and of casemate walls on the other), in the structure of gates and in various weapons (such as battering rams) – demonstrates the continuous development of offensive and defensive techniques in city warfare.[6] In general, the discussion

5 See, for example, Crété, *La vie quotidienne à La Rochelle au temps du grand siège: 1627–1628*, and especially references therein, 323–329; Pavlov, *Leningrad 1941: The Blockade.*

6 See, for example, Yadin, *Art of Warfare*, 16–24, 50–57, 65–71, 90–100, 253–254, 261–262, 267–270, 287–290, 313–328.

here has been limited to a description of the basic technical phenomena and their principal features. However, it must be remembered that every battle over a city had specific characteristics, stemming from local topographical and technical conditions.

The examination of siege warfare is based on a wide range of facts from various countries over an extended period of time, the detailed discussion of which may at times obscure the overall picture. To avoid this and to ensure that the information "trees" do not obscure the "forest" of the subject discussed, the primary subject matter has been elaborated in the text, while the technical details have been confined to footnotes. As stated above, it is not the purpose of this book to describe any particular siege; rather, our goal is to identify the characteristics of siege warfare as a phenomenon, as revealed in extant sources from the ancient Near East.[7] Occasionally, a certain aspect may pertain to several areas. In order to avoid repetition in such cases, discussion of the relevant aspect is limited to a single section. A reader interested in the full range of data on a particular subject is referred to the index.[8]

In some of the discussions dealing with military technology, a basic methodological problem emerged: at times, the exact meaning of technical terms appearing in written sources is uncertain, while at other times we lack the technical terms for items clearly depicted in ancient art.[9] Our technical discussions are primarily based on concrete data – from the arts, and at times from military reality as well. Thus, even if the linguistic definition of certain technical terms turns out to be erroneous, this will not affect the underlying military logic of the

7 This approach is primarily derived from the awareness that our knowledge of military events in ancient times is so limited as to make it impossible to reconstruct any particular event in detail. See Ephʿal (n. 1 above), 106.

8 Thus, for example, the chopping down of trees during a siege is discussed both in the chapter concerning sources (below, p. 15) and in the discussion of negotiations (below, p. 54).

9 For example, on the one hand, there is no certainty in the interpretation of the terms *pilšu*, סֹלְלָה, or *kalbānātu*, and דָּיֵק = *dāiqu*; on the other hand, we are not sufficiently familiar with all of the terms used for the battering ram and its various parts, or for the ladders which are clearly depicted in Assyrian and Egyptian reliefs.

discussion (although it may, of course, affect the quantity of written evidence regarding the use of one specific method or another; see, for example, the discussions of tunnels and assault ladders, in Chapter 3).

Every siege is filled with tumultuous drama both on the personal and the collective levels; all the more so a book in which numerous cases of sieges have been assembled and their various aspects described. Accustomed to the factual approach of scientific writing, I have chosen to present the manifestations of siege in a concrete, matter-of-fact style. Nevertheless, it is my hope that the sensitive and imaginative reader will be able to discern the personal aspect hidden behind the dry language, that he will hear the sounds of victory and defeat, of enthusiasm and pain which accompanied them, and that he will sense the determination and continued strain experienced by all those involved in a siege – attackers and defenders, combatants and non-combatants alike.

CHAPTER TWO

SOURCES

The phenomenon of siege warfare in the ancient Near East and in neighboring regions is reflected in various fields, described in a variety of different extant sources – in written documents, in the plastic arts and in archaeological findings. Some of these sources reflect a military, economic, legal and, to a certain extent, also social reality; others relate to the spiritual world of the participants, their ideologies and their literary and linguistic heritage. Archaeological remnants, non-literary documents and technical descriptions depicted in artistic sources probably bear witness to actual events. In contrast, various types of literary sources raise the question whether they reflect reality (not including reference to miraculous events, which are totally irrational), to what extent, and if so, what kind of reality is it? Is it a concrete, factual reality that took place at a particular time and place, or a historical reality whose details correspond to the conditions of a given situation, yet may not have actually occurred?

A detailed description of the sources at our disposal and their characteristics is beyond the scope of this book. A brief survey of the various types of sources will suffice, noting characteristics that directly pertain to the study of siege warfare in the ancient Near East.

I. LITERARY SOURCES

The literary sources used in this book include hundreds of literary works written over a period of two thousand years or more, from the Sumerian "Curse of Akkad" and the "Lamentation on the Destruction of Ur," through various books of the Hebrew Bible and the royal inscriptions of the ancient Near East, to the writings of Greek historians on Alexander the Great. These works are varied, both in genre (personal reports, chronicles, stories, speeches, lamentations and law) and in the authors' goals (political propaganda, ideological-religious, didactic or practical writings). The nature of the information that can be gleaned from these works as historical sources is inevitably related to the authors' aims and to the nature of the genre.

1. Royal Inscriptions

These inscriptions originated in the royal courts and in related scribal circles. They were intended to commemorate royal projects and therefore naturally lack any mention of failures of the kings on whose behalf they were written. Ideological and propagandistic tendencies can be discerned in the style of these inscriptions, in the details described therein and in the manner of the descriptions. This may raise doubts as to the historical reliability of the events described.[1] Among the inscriptions relating to the siege or conquest of cities, we find Mesopotamian royal inscriptions – from the Old Babylonian period (Dadusha, king of Eshnunna) to the final generation of the Assyrian empire (Ashurbanipal) – Egyptian royal inscriptions – of Thutmosis III and Pi(ᶜankh)y – and Hittite royal inscriptions, listing the kings' accomplishments in bringing about the surrender and defeat of various cities. There are also royal inscriptions of kings of the western Fertile Crescent, which commemorate their successes in withstanding forces stronger than their own (Mesha, king of Moab and Zakkur, king of Hamath and Luᶜash).[2]

Most relevant royal inscriptions are Assyrian, particularly from the Neo-Assyrian empire (9th–the second half of the 7th centuries BC). These tend

1 On the tendentious characteristics of royal historiography and its distance from military and diplomatic reality, see Liverani, *Prestige and Interest*, esp. 115-202. Liverani's book focuses primarily on Egyptian and Hittite royal literature of the 16th to 12th centuries BC, but many of his distinctions are valid as well to royal writing in later periods in the ancient Near East.

2 A siege is also mentioned in a fragment of the Tel Dan Aramaic inscription, line 13, see Biran and Naveh, *IEJ* 43 (1993), 87. Due to the fragmentary condition of the inscription, it is impossible to determine the circumstances of the siege described.

to be schematically formulated: "…I laid siege to the city [name of city], I conquered it and destroyed it / burned it down." The means and methods used in the conquest of particular cities are rarely enumerated, but even the more detailed descriptions seem to be literary conventions (albeit of a higher level than of the standard royal inscriptions).[3]

Few Egyptian royal inscriptions relate to siege warfare (there are only two, I believe: the Pi(ʿankh)y inscription and the "Annals" of Thutmosis III). Nevertheless, compared to the Akkadian royal inscriptions, the Pi(ʿankh)y inscription includes a wider range of data regarding siege warfare.

2. The Historical Story

The historical accounts of various types are replete with many details that are the result of ideological, artistic or technical-literary phenomena and they do not necessarily record events as they occurred. Our use of such records is based on the assumption that the accounts were meant to be accepted by the audience for whom they were composed, and therefore reflect military techniques or political, social and economic realities operative at the time of composition. Nevertheless, there is no certainty that these events took

3 Thus, for example, the Sennacherib inscriptions relate that the cities of King Hezekiah of Judah were besieged "by constructing of siege ramps, by bringing up battering rams, by an infantry attack, and by tunnels and breaches" (OIP II 32–33 iii 18–23). Similarly, an Esarhaddon inscription states that Memphis, the capital of Lower Egypt, was captured by the use of mines, breaches and other means of breakthrough in "half a day" (Borger, *Asarh.*, 99 rev. 41–43). If we accept the data of the Babylonian Chronicle as accurate (Grayson, *Chron.* 1 iv 23–26) – namely, that the battles between the armies of Assyria and Egypt took place on the 3rd, 16th, and 18th of Tammuz, 671 BC, and that Memphis fell on the 22nd of that month – it becomes clear that there was no real siege against the city and that Esarhaddon had no time to prepare the means of breach listed in the inscription. It seems more likely that Esarhaddon's scribe adopted, in this inscription, the formula that had been set in the Sennacherib inscriptions in 700 BC, which reflected the reality of his own war in Judah.
Reference to the conquest of a city within a very short period of time is a common literary motif in the ancient Near East. On this motif in the Assyrian royal inscriptions, see Tadmor, in *Assyrian Royal Inscriptions: New Horizons in Literary, Ideological and Historical Analysis*, 17–18, and compare the biblical account of the Israelite conquest of the cities of southern Judaea following a siege – as implied by the expression חנה על ("laid camp against") – ביום ההוא ("on that day"; Josh. 10:34–35), or ביום השני ("on the second day"; *ibid.* vv. 31–32).

place as described. Since this book is concerned with the typology of siege rather than with the description of specific realities of a given time or place, it makes no difference, for our purposes, whether the data in literary sources reflect the details of a particular historical experience. Even with respect to events generally considered historical,[4] one must distinguish the basic information (such as the occurrence of a siege and its outcome) from the various details present in the written text (such as miracles, folklore and ideology) expressed by the narrator or by the various characters.

This approach is exemplified with the biblical story of Elisha and the siege of Samaria (2 Kgs. 6:24–7:20). Seemingly, we may accept the historical accuracy of the report of a siege imposed by King Ben-Hadad of Damascus on Samaria and the subsequent failure of that siege. However, we must separate secondary plots from this basic information (if it is indeed historical) in light of their particular characteristics:

(i) In this passage, the narrator explains the flight of the Aramean camp by an act of God that caused the army to hear "a sound of chariots, a sound of horses, the sound of a great host, so they said to one another, 'The King of Israel has hired the kings of the Hittites and the kings of Egypt against us to attack us'" (7:6–7). This is clearly a miracle, although the practice of hiring foreign armies in the case of military inferiority is well documented in ancient Near Eastern sources.[5]

(ii) The story of the women who agreed to eat their offspring and the fulfillment of this agreement (6:26–29) is imbued with the motif of extreme cannibalism – parents eating their own children. This motif is part of the famine *topos*, and is quite common in the literary tradition of the ancient Near East.[6]

4 These are contrasted by traditions that clearly cannot be taken as accounts of historical events, in light of literary analysis or archaeological findings. These traditions include the stories concerning the Israelite conquest of Jericho and Bethel and at least one of the two traditions concerning the liberation of Jeremiah by Nebuzaradan, the captain of the guard of Nebuchadnezzar, king of Babylon (Jer. 39:11–14; 40:1–5). See below, pp. 165–167.

5 See 1 Kgs. 15:18–20; 2 Kgs. 16: 7–9 (cf. 2 Chron. 28:21); the Kilamuwa inscription (*KAI* 24) ll. 7–8; and also Liverani, *Mesopotamia* 17 (1982), 61–63.

6 See below, pp. 61–62.

(iii) The extreme prices of food such as grains, both inflated and deflated, noted in the story of the king's officer who mocked Elisha (7:1–2, 16–20; cf. 6:25), are taken from well-known formulae. These formulae are part of the famine *topos* on the one hand and the abundance *topos* on the other, both common in the literary tradition of the ancient Near East.[7]

(iv) The incident of the lepers (7:3–10) is a pure folk tale.

Although these secondary stories are not directly connected to the Aramean siege of Samaria, they can be used to discuss its literary aspects and contribute details on the military, political and social reality of the siege.[8] Some of these details may be incorporated into the description of siege as a phenomenon, particularly during the 8th and 7th centuries BC, during which the legend of Elisha's role in the siege of Samaria took shape.

In the beginning of his *History of the Peloponnesian War*, Thucydides describes his method of writing:

> As to the speeches that were made by different men, either when they were about to begin the war or when they were already engaged therein, it has been difficult to recall with strict accuracy the words actually spoken, both for me as regards that which I myself heard, and for those who, from various other sources, have brought me reports. Therefore the speeches are given in the language in which, as it seemed to me, the several speakers would express, on the subjects under consideration, the sentiments most befitting the occasion, though at the same time I have adhered as closely as possible to the general sense of what was actually said. (1.22)

As noted earlier, some of the details cited in historical narratives are fictional *ab initio*; most of these texts were written long after the events described and far from the original locations, and were not intended to be accurate historical documentation. Thucydides' comments are even more apparent with respect to the characterization of the texts assigned to the various figures in this literary genre. These are to be understood as no more

7 See below, pp. 124–127; cf. *ibid.*, pp. 129–135.

8 In addition to the references in the two previous footnotes, see below, pp. 144, n. 83.

than *literary materials*, the products of the understanding and imagination of their authors. Hence, the texts ascribed to the various characters in the largely fictional story of Elisha at the siege of Samaria (including the words of the woman who cried out in the controversy over the dead child and the living child, as well as those of Elisha, the king, the lepers and the king's servant) are presumed to be literary material and not text actually spoken by the various participants. Similarly, Rabshakeh's statements to the representatives of King Hezekiah of Judah and the people upon the wall (2 Kgs. 18:19–25, 28–35 // Isa. 36:4–10, 14–20) are no more than arguments chosen by the narrator, suited to his perceptions and to those of his Judean audience. Thus, despite a certain linguistic and stylistic similarity between Rabshakeh's speeches and the inscriptions of Sennacherib, king of Assyria,[9] it is unlikely that these statements can provide any information on the methods of persuasion employed by the Assyrian army or its statesmen. Moreover, it is doubtful that Rabshakeh's statements (2 Kgs. 18: 22, 25 // Isa. 36:7, 10) can be taken as evidence for the Assyrian official's knowledge of the cultic reality in the kingdom of Judah or of the religious beliefs of its inhabitants.[10] Even the famous dialogue between the leaders of Melos and the Athenian representatives seeking their surrender (Thucydides 5.85–113), a dialogue filled with wisdom and experience of political and human thought and invoked as an example of diplomatic negotiation, is no more than a literary creation, and it is unlikely that the historical event took place in the manner described in the text.[11]

9 See Cohen, *IOS* 9 (1979), 32–48.

10 On the assumption that Rabshakeh's origin was Aramean, or even Israelite, see Tadmor, "Rabshakeh," *Enc. Miqr*. 7. 323–325. The argument that there should be no expectations for God's salvation, since He wished to punish Judah for Hezekiah's sinful religious reforms (see 2 Kgs. 18:22 // Isa. 36:7; and cf. 2 Kgs. 18:25 // Isa. 36:10), reflects a negative view of these reforms among some circles in Judah. There is no doubt that the criticism of Hezekiah's religious policy, articulated by Rabshakeh in the name of the Assyrian king, was ascribed to an Assyrian official rather than to a Judean prophet as part of the rhetoric of the story and need not be related to historical reality. See below, p. 50, n. 41.

11 On the shaping of this dialogue in light of political developments following the conquest of Melos by Athens, and in light of the attitudes that guided Thucydides rather than the participants in the negotiations, see de Romilly, *Thucydides and Athenian Imperialism*, 273–310.

These reservations negate almost entirely the possibility that texts ascribed to various characters could be regarded as factual history of specific events. Nevertheless, these texts can be used for the examination and creation of a typology of historical phenomena in general. They have, therefore, been included in our study, as expressions of the siege phenomenon.

3. Greek Historians

The wars between the Greeks and Persians brought two military traditions together; the Persians brought the accumulated military experience of the ancient Near East, and faced techniques of warfare as well as social and political phenomena of military significance that originated in Greece and in the West. Despite the difference between the two traditions, the mutual influences characteristic of military activities make both significant in illuminating various aspects of the period discussed in this book, found in the writings of classical authors, but absent from texts of the ancient Near East. Therefore, reference is made to the writings of Herodotus and Thucydides (composed in the second half of the fifth century BC; Herodotus preceded Thucydides by several decades). Herodotus' work incorporates information concerning military activities of kingdoms of the ancient Near East, particularly those of the Persian Empire, while Thucydides' work focuses primarily on the first twenty years of the Peloponnesian War (431-411 BC), during which siege tactics played an important role. Nevertheless, our discussion remains limited to those aspects (technical or social) which shed light on the siege phenomenon in the ancient Near East; the entire complex of siege warfare in Greece and in western Asia Minor during the ancient period will not be discussed here.

The latest events in the period covered by this book are Alexander the Great's wars against Thebes, Halicarnassus, Tyre and Gaza. The historians who chronicled these wars – Diodorus (1st century BC), Curtius Rufus (1st century CE) and Arrian (2nd century CE) – lived in a period much later than the chronological framework of our discussion; yet, their works are included due to the great importance of Alexander's siege battles. The descriptions of the battles of Tyre and Gaza by these three authors display differences in technical detail, but include similarities indicative of a common early source. A full description of the relations among these accounts would require a separate examination of each and every detail.[12]

Other authors, such as Polybius (2nd century BC), Onasander (1st century CE) and Pausanias (2nd century CE), are mentioned in passing, where their writings contain technical details important for the understanding of the ancient data.

4. Biblical Oracles

The three larger books of the Latter Prophets – Isaiah, Jeremiah and Ezekiel – reflect a unique viewpoint on the experience of the siege of Jerusalem. These prophets did not belong to the circles of royal scribes writing literary propaganda; rather, their words essentially express their own personal perceptions of reality. Hence, these oracles express political and social aspects that differ from those in the other extant sources. Concurrently, the oracles contain siege-related *topoi* (including descriptions of famine, disease and similar hardships) that belong to the common literary tradition of the ancient Near East, and do not necessarily reflect specific historical circumstances.[13]

5. Laws

The laws of warfare in the book of Deuteronomy (20; 21:10–14; 23:10–15; 24:5) include several laws specifically related to siege warfare (20:10–20).[14] According to Deuteronomy 20:10–14, before one lays siege to a city, one must call upon it in peace, and if it opens its gates, one must make

12 On this relationship and the assumption that Diodorus and Curtius Rufus relied on a common source (Cleitarchus, 3rd century BC), see Hammond, *Three Historians of Alexander the Great*, 42, 50–51, 118–125, 166–169.

13 This common literary heritage is reflected, for example, in identical images that appear in the biblical oracles and in the imprecations detailed in the sanction clauses of political treaties in the ancient Near East. See Hillers, *Treaty-Curses and the Old Testament Prophets*.

14 On the laws of warfare in the Book of Deuteronomy, including the laws of siege warfare, their origins, purpose and positive character, see Rofé, *JSOT* 32 (1985), 23–44.

peace with it.[15] It is clear from verses 15–18 that verses 10–14 only apply to "the towns that lie very far from you," whereas it is forbidden to arrive at any kind of arrangement with the indigenous population of the land of Canaan and one is obligated to destroy them. According to verses 19–20, it is forbidden to destroy fruit-bearing trees during a siege of a city.

Of these, only the laws in verses 10–14 are to be understood as a practical guideline of universal applicability: the conquest of a city requires considerable effort, and it is therefore preferable to attempt to convince the inhabitants to surrender by means of negotiations (whether through friendly means or by intimidation).[16] By contrast, it is debatable whether the two latter laws were ever applied in practice. On the one hand, verses 15–18 are not part of the ancient law found in verses 10–14, but belong to a later literary level. These verses relate to the notion of the proscription (חרם), and the terms "the remaining land" and "the promised land" stem from a retrospective program formed in the last generations of the age of the monarchy, and were certainly not in use during the Israelite conquest and settlement of the land. On the other hand, verses 19–20 are characterized by humanitarian idealism and lack practical reality. Elisha's command to the kings of Israel and Judah in their campaign against Moab, delivered in the name of God: "You shall conquer every fortified town and every splendid city, you shall fell every good tree and stop up every spring, and every fertile field you shall ruin with stones" (2 Kgs. 3:19; cf. verse 25), is similar to other reports of the behavior of warring armies[17] and seems closer to the Israelite military reality. We may therefore view the laws in Deuteronomy 20:15–20 as theoretical and never carried out in practice.

15 On the nature of this peace, i.e., surrender and becoming tributary, see below, pp. 46–48.

16 Compare, below, the queries to the gods and pp. 43–54. See also the letter to Zimri-Lim, king of Mari from Iddiyatum, whose army was stationed in the vicinity of Karana: ARM II 42(=ARM XXVI 518):8: *ištu ā[lam] šuatu ilwû salī[mam] iššīšumma*, "After he surrounded the city of Asna he called upon it for peace."

17 Compare, 2 Sam. 11:1 (and its parallel in 1 Chron. 20:1): "…David sent Joab with his officers and all Israel with him, and they devastated Ammon and besieged Rabbah." On the felling of trees and other destructive measures during fighting over cities, see below, Chapter 3.1 (1.ii); cf. Rofé (above, n. 14), n. 31.

6. Practical Military Literature

Another source, quite different in character from the literary-historical works on which we rely in this book, is Aeneas the Tactician's treatise on the defense of fortified positions from the mid-fourth century BC.[18] This treatise, the only surviving text by this author on military matters (his other works are known by name only),[19] is one of a genre of practical military handbooks that began to appear in Greece from the end of the fifth century BC. This is the earliest of its type to deal with the topic of siege, and consists of directives to the defenders (indirect evidence suggests that another work by the same author, *Poliorcetica*, was dedicated to siegecraft).[20]

A considerable part of Aeneas' discussion and the examples he uses reflect the social and political conditions of Greek cities in the fourth century BC and the struggles between the oligarchic parties and the popular parties found there; it is thus very doubtful that they correspond to the predominant social and political reality in the ancient Near East. However, a good part of his instructions focus on technical matters suitable to any besieged city irrespective of its form of government. Among the subjects covered in his writings are directions on pre-designated signals between the people in the besieged city and their comrades outside (Chapter 4), on passwords (Chapter 24), on means of secret communications and methods of preventing them (Chapter 31), on dealing with the armies of allies and with mercenaries within the besieged city (Chapters 12–13), on the treatment of relatives of hostages who are likely to cooperate with the enemy or to rebel when they see their loved ones tortured (Chapter 10.23–25), on the supervision of the local population to prevent rebellion (Chapters 2 and 11), on the ways to prevent panic (Chapter 27), on placing

18 The work of Aeneas the Tactician has appeared in three different translations: *On the Defense of Fortified Positions* (trans. the Illinois Greek Club, 1923); *Aeneas on Siegecraft* (trans. Hunter, 1927); *How to Survive Under Siege* (trans. Whitehead, 1990).

19 See Aeneas (1923), 8–10.

20 Another treatise worth mentioning in this context, although it departs from the chronological framework set here, is the treatise on technology by Philo of Byzantium (ca. 200 BC). Book 8 of this work (*Poliorcetica*) is devoted to the construction of offensive and defensive works and other measures to be taken in the event of siege; see Diels and Schramm, *Excerpte aus Philons Mechanik B. VII und VIII.*

guards on the wall and their supervision (Chapter 22), on the organization of combatants within the besieged city in routine times (Chapter 1) and during enemy attacks, to better maintain fresh troops and reserves (Chapter 38), on methods of locating tunnels under the walls and preventing their use (Chapter 37), on the repulsion of assault ladders from scaling the wall (Chapter 36), on setting enemy machinery on fire and extinguishing fires caused by the enemy (Chapters 33–35), on the execution of night raids on the enemy and the organized return of raiders to the city (Chapter 23) and on sorties through the city gate and withdrawal within it, while breaking off contact with the pursuing enemy (Chapter 39). As noted, these were practical guidelines, independent of the form of regime. It is, therefore, likely that they were also practiced throughout the ancient Near East. The great variety of these topics reflects a range of military and social aspects that rarely find expression in the extant information on the ancient Near East, primarily due to the nature of our sources from this region.

II. NON-LITERARY DOCUMENTS

Unlike the literary sources, which are problematic in terms of objectivity and historical reliability, non-literary documents are of special significance, first and foremost because they are "innocent witnesses" and are close to the events, both chronologically and in subject matter. Their significance lies in reflecting not only military technology but different aspects of siege in its broadest sense. These sources include letters, siege documents, mathematical texts and others.

1. Letters

Letters from various periods are mentioned in this study, beginning with letters from Mari and Shemshara, written during the Old Babylonian period, and ending with letters from the time of the Assyrian Empire. These letters provide direct accounts (often first-hand but occasionally containing reports of second-hand information); yet, it is often hard to determine the exact circumstances under which these accounts were written, which were clear to the author and recipient. The style of these letters is not literary, and their exact meaning may be difficult to decipher due to the appearance of unique syntactic and grammatical phenomena.

2. Siege Documents

This type includes legal documents containing formulae explicitly noting their composition in time of siege, which reflect the unique economic and social reality of this emergency situation. They document economic and legal activities that took place during sieges such as enslavement in exchange for food, the sale of children and real estate at a low price, loans under special conditions and possibly even the sale of rights to prebends delivered to officients in Babylonian temples. Siege documents are known from the city of Emar in northern Syria (13th–12th century BC) and from seventh century BC Babylonia.

3. A Contract

One of our sources is a contract between the authorities of Idalium, Cyprus, and a physician, assuring free medical treatment for anyone injured in battle within one year of the signing of the contract.

4. Divinatory Literature

Documents dealing with divination are found within the entire chronological framework of this book, from the end of the Old Babylonian period to the Seleucid period. There are two different types of document containing data pertinent to this study:

(a) The divinatory literature of Mesopotamia and countries under its cultural influence (such as Elam and the Hittite kingdom) includes a series of observations, some real and some theoretical, that are the result of analogies and comparisons, and include data for divination.[21] These series are composed of conditional clauses each beginning with the enumeration of various signs in accordance with the particular technique of divination and ending with the interpretation of these signs. The interpretations include military activity, some relating to pitched battle, while most of them relate

21 On divination within the Mesopotamian cultural sphere, see Oppenheim, *Ancient Mesopotamia*, 206–227; *idem*., "Divination," *Enc. Miqr*. 6. 421–426.

to siege and its manifestations.[22] For example: "If an eclipse occurs on the 16th of Shebat, famine will seize the city, the wall will be destroyed… the king will be captured" (*ACh* Sin XXXIII 76).[23] The signs included in the series were applicable at all times, and their interpretation valid for all cities.

(b) *Queries to the God(s)*: Another way to obtain knowledge of the future was by directing a query to the oracular gods, Shamash and Adad. There are hundreds of extant documents from the reigns of Esarhaddon and Ashurbanipal, formulated as queries to Shamash. The questions recorded therein on behalf of the two Assyrian kings were intended to verify reports they had received and to provide them with information concerning the future – based on inspection of livers of rams slaughtered for the occasion – before making decisions in diplomacy, military matters, imperial administration or personal health.[24] A number of these documents concerning military activity relate to the fate anticipated by cities that may be subjected to siege by the enemies of Assyria or to those enemy cities to which Esarhaddon or Ashurbanipal intended to lay siege. A similar type of document are the *tamītu* inscriptions, in which Babylonian kings of the Old Babylonian period addressed Shamash and Adad concerning their own activities planned for the near future, including military activities.[25] The

22 On omens related to war and its phenomena, including siege, see Oppenheim, *Orientalia* 5 (1936), 199–228, esp. 208–211. On queries related to siege, see ARM XXVI 121:9 ff.

23 For the combination of episodes detailed in this interpretation, cf. Jer. 52:6–9 (//2 Kgs. 25: 3–6): "By the ninth day of the fourth month, the famine was severe in the city and there was no food left for the common people. Then (the wall of) the city was breached… the Chaldaean troops pursued the king, and they overtook Zedekiah in the steppes of Jericho, as all his troops dispersed. They captured the king and brought him before the king of Babylon..."

24 For the first publications of the queries to Shamash, see: Knudtzon, *Assyrische Gebete an den Sonnengott*; Klauber, *Politisch-religiöse Texte aus der Sargonidenzeit*. All the texts published by Knudtzon and Klauber have been re-edited, with the addition of a number of new texts, by Starr, *Queries to the Sungod* (=SAA IV). Here they are quoted according to Starr's edition. On the characteristics of the queries to Shamash and on divination in the service of the kings of Assyria based upon inspection of the entrails of animals, see Starr, *ibid*., xiii–xxxv.

25 On the nature of the *tamītu* documents and the chronological and stylistic relations

queries to the god(s) are distinguished from the omen series mentioned above because, among other reasons, the questions refer to specific events or personalities, and the practical applicability of the divination is limited in time (7 to 100 days in the queries to Shamash, up to one year in the *tamītu* inscriptions). These documents seem to reflect a straightforwardness similar to that of a patient consulting his physician or a person who turns to an omniscient god, in the belief that it is impossible to hide anything from him.[26]

The queries relating to siege enumerate the ways and means of capturing a city, some of which are defined in concrete technical terms and some in abstract terms. Each query mentions a large number of methods. Thus, for example, SAA IV 102 lists twelve ways of capturing a city; according to the reconstruction of the editor, Starr, the missing sections contained about eight additional options. To ascertain that the query encompasses all possibilities for capturing a city, the detailed questions in the queries to Shamash conclude with the general question "or (will the city be conquered) through any ruse?"

The various methods of capturing cities are listed here without distinction between the two groups of queries, due to the clear similarity between the groups in the nature of the requested information and the detailed format. This provides an almost complete "catalogue" of methods (the following text is, of course, eclectic):[27]

> Shamash, great lord, give me a firm positive answer to what I am asking you! From this day, the 3rd day of this month, the month Iyyar, to the 11th day of the month Ab of this year, for these 100 days and nights, the term stipulated for the performance of (this) extispicy

---

between them and the queries to Shamash, see Lambert, in *La divination en Mésopotamie ancienne et dans les régions voisines*, 119–123; *idem*, in *Oracles et prophéties dans l'antiquité*, 85–98; Starr, *ibid.*, xxix. The *tamītu* inscriptions quoted here are to be published by W. G. Lambert. Reference to them is made on the basis of their citations in *CAD*.

26 On extispicy in Mari in connection with siege, see ARM XXVI 121, 169 (in the latter document, the divination is valid for one month).

27 As the basis for this eclectic text, I used SAA IV 43.

– within this stipulated term, will Kashtaritu (city lord of Karkashshi) with his troops, or the troops of the Cimmerians, or the troops of the Medes, or the troops of the Manneans, or any other enemy, strive and plan? Will they capture the city of Kishassu through friendliness and peaceful negotiations[28] (which result in) a treaty invoking the names of a god or a goddess,[29] or by fearfulness and pressure?[30] Or by force?[31] By a blockade,[32] i.e., by famine, hunger and want?[33] Or by thirst?[34] By waging war and breaking through,[35] i.e., by siege towers[36] and (scaling) ladders,[37] or by (piling up) siege ramps[38] and other means of breakthrough,[39] or by battering rams,[40] certain siege engines(?) and a powerful weapon?[41] By a tunnel, or a breach?[42] Or by water that softens (bricks),[43] or by fire?[44] Or by negligence (of the

28 *ina pî ṭābi u salīm/silīm ṭubbāti* (SAA IV 30:⌜6⌝, 43:9, 44:⌜10⌝, 267:⌜5⌝, rev. 10, 81-2-4, 209:8); *ina dibbī ṭābūti* (SAA IV 101:5).

29 *adê zakār šumi ili u ištāri* (SAA IV 43:9).

30 *puluḫtu* (*SAA* IV 29:5'); *si'ūtu* (harassment?, SAA IV 43:6, 44:8, 102:17). Compare also *uššušu* (*IM* 67692:5).

31 *emūqu* (IM 67692:52); *danānu* (SAA IV 43:6, 63:6, 102:6); cf. also *rittu* (IM 67692:55).

32 *esēru* (IM 67692:52).

33 *bubūtu* (SAA IV 29:⌜4⌝, 30:7, 44:9; IM 67692:59; 81-2-4, 209); *ḫušaḫḫu* (SAA IV 29:4; IM 67692:59; 81-2-4, 209 obv.? 7); *sunqu* (SAA IV 29:4; IM 67692:59).

34 *ṣūmu* (SAA IV 102:6; IM 67692:59).

35 *ina epēš kakki qabli u tāḫāzi* (SAA IV 31:⌜6-7⌝, 43:7, 44:8, 63:⌜5⌝, 102:⌜3⌝, 267:5, rev. 10; *ina kakki danni* (SAA IV 102:5').

36 *dimtu* (IM 67692:53).

37 *simmiltu* (SAA IV 30:8, 43:7, 44:9, 102:4'; IM 67692:55; 81-2-4, 209:3).

38 *arammu* (SAA IV 29:3', 43:8, 101:6', 102:⌜4⌝); *eperī šapāku* (IM 67692:53).

39 *kalbānātu* (IM 67692:55; 81-2-4, 209 obv? ⌜3⌝; K 3467 iii 12).

40 *šubû* (SAA IV 43:⌜8⌝, 44:10, 63:7); *āšibu* (IM 67692:54). On the breaching of the wall by hand (with the help of a dagger), cf. *ina patar pāliši* (IM 67692:58).

41 *ina kakki danni* (SAA IV 102:5).

42 *pilšu* (SAA IV 31:7, 43:⌜7⌝, 44:9, 102:⌜4⌝); *neptû* (IM 67692:53); *niksu* (SAA IV 31:7; 43:⌜7⌝); *nikis dūri* (IM 67692:56).

43 *mê maḫāḫi* (SAA IV 102:5'). Cf. IM 67692:60 (in CAD M/I 49b, *maḫāḫu*): "Will the city be conquered by [*iṭ-ṭ*]*e-e maḫāḫi*, by softening bitumen?" The danger of flooding during a siege was expected for cities located on flat land adjacent to rivers. Thus, it is reported by Diodorus (15.12) that the city of Mantinea, which for a protracted time withstood the siege of the Lacedaemonians, surrendered in winter 385 BC after

defenders),[45] or through lack (of warriors in the city)?[46] Or through insurrection, revolt,[47] rebellion[48] and treason?[49] Or through any ruse of capturing a city?[50] Will they enter that city, Kishassu? Will they conquer that city, Kishassu? Will it be delivered to them? Your great divinity knows it. Is the capture of that city, Kishassu, by any enemy,

---

its besiegers redirected the waters of the high river that flowed inside it and in its close environs, causing the collapse of its houses. The redirecting of a canal or a river flowing in flat terrain was a feasible military activity. According to Herodotus 1.191, Cyrus redirected the waters of the Euphrates that flowed within the city of Babylon, thereby enabling his soldiers to enter the city. (Compare, similarly, Thucydides 1.109, on the redirecting of a canal of the Nile and its drying out during the siege of an army, commanded by Megabyxus, upon the Athenian naval force that came to Egypt to help Inaros, who was struggling against Persian rule.) A legendary tradition cited by Diodorus (2.26.8 – 27.3) states that the besieged city of Nineveh fell after the waters of the Euphrates (Diodorus is probably referring here to the Tigris) rose and caused its walls to fall. For the conjecture that this alludes to the redirecting of its waters by the attackers (the Babylonians and the Medes in 612 BC), and does not refer to flooding due to the natural rise of the river in wake of intense rainfall, see Scurlock, *Historia* 39 (1990), 382-384.

44 *išātu* (NE): This term appears in an unpublished *tamītu* text together with *pilšu*. In Assyrian reliefs depicting battles over cities, the attacking soldiers are shown setting fire to the gates of the city with their torches; see Yadin, *Art of Warfare*, 421, 425, 462. The literary sources mention setting fire to piles of wood that were thrown upon the wall and even beyond it, within the besieged city or next to the gate (Thucydides 2.77; compare Judg. 9:46-52; cf. below, p. 89, n. 164), as well as setting fires within the city by fiery arrows shot over the wall (as was done by the army of Xerxes in the battle for Athens, Herodotus 8.52; compare the rain of flaming arrows upon the enemies of Ashurnasirpal II, king of Assyria, RIMA 2 225:22). In this light, perhaps the expression *šiltaḫ girri*, in the list of equipment of Gedaliah son of Rahimel, who joined the service of the Persian cavalry in 423 BC (*UCP* 9/III 275:9), ought to be seen as a technical term for fiery arrows. For other explanations of this term, see Ebeling, ZA 50 (1952) 206-207.

45 *šēṭūtu* (SAA IV 102:7').

46 *mēkûtu* (SAA IV 29:2', 30:7, 102:7'); *mēkûtu ša ummanāti ša libbi āli* (SAA IV 31:8).

47 *bartu* (SAA IV 63:8; IM 67692:49); *sīḫu* (SAA IV 63:8; IM 67692:49).

48 *nabalkattu* (SAA IV 43:7; IM 67692:56; 81-2-4, 209 obv.? 3); *gabaraḫḫu* (IM 67692:57); *tēšû* (IM 67692:67).

49 *sartu* (IM 67692:50).

50 *mimma šipir nikilti ša ṣabāt āli mal bašû* (SAA IV 30:⌜10⌝, 43:10, 44:11, 102:8', 267 rev. 11).

> from this day to the day of my stipulated term, decreed and confirmed in a favorable case, by the command of your great divinity, Shamash, great lord? Will he who can see, see it, he who can hear, hear it?… [instructions regarding signs to be read in the liver] … Be present in this ram, place (in it) a firm positive answer, favorable designs, favorable, propitious omens by the oracular command of your great divinity, and may I see (them). May (this query) go to your great divinity, O Shamash, great lord, and may an oracle be given as an answer.

This list detailing dozens of methods is, as mentioned above, an almost complete catalogue of the means of conquering a city, and will serve as an outline for our discussion of military aspects in the next chapter.

5. Mathematical Texts

Babylonian mathematical texts include various engineering exercises calculating the dimensions of walls and siege ramps near them (based on which the quantities of materials and labor required to lay down those ramps can be calculated).[51]

51 See, for example, BM 85194 rev. ii 7–33, in Neugebauer, *Mathematische Keilschrifttexte* I, 149; cf. *ibid.*, 182–186. The war-like context of these calculations also follows from the text, which states that one is speaking of "a city [that is] an enemy of Marduk" (*ālu nakir Marduk*, rev. ii 7, 22). An example of quantitative data concerning the fortifications of a city in the context of an attack occurs in an inscription engraved upon the back of the stone monument of King Naram-Sin of Akkad, which stood in the temple of Sin in Ur (the monument itself did not survive; all that is extant is a copy of the inscription and a verbal description of the relief, prepared by an Old Babylonian scribe). The monument depicted the conquest of the city of Armanum. The description of the city's fortifications records the heights of three walls that surrounded it, the length of the slopes between them and the sum total of these numbers; see Foster, *JANES* 14 (1982), 27–36 (Foster's explanations regarding the combination of the quantitative data seems somewhat anomalous); *idem, Before the Muses* I, 53.

Fig. 1: The conquest of Gezer. A relief from the palace of Tiglath-pileser III in Calah (Nimrud)
From Layard, *The Monuments of Nineveh*, I, Pl. 62

III. ARTISTIC REPRESENTATIONS OF SIEGE WARFARE

The primary, and almost only, source of information concerning the various activities conducted during siege and the technical aspects and devices used in siege warfare are visual portrayals of combat and siege on reliefs in Assyrian royal palaces and Egyptian temples, and on the bronze strips which strengthened and decorated the gates of the city Imgur-Enlil (Tell Balawat) and, to a lesser extent, in frescoes discovered in Egyptian tombs.[52]

52 For analysis of the military scenes in these reliefs, see in particular Yadin, *Art of Warfare* and references therein.

By way of these images, we know of the techniques of breakthrough (scaling walls, tunneling under walls, breaching gates or walls), of the means used during these operations, and of their development over time;[53] we learn of the use of fire in siege and of the various stages in the battle over a city, from the beginning of the siege of a city to its conquest and destruction, and the deportation of its inhabitants. Examination of the reliefs enables us to draw certain conclusions concerning the reality of military technology, based not only on what is depicted, but also on what is absent. Thus, for example, we may deduce from the absence of battering rams in the breakthrough scenes in the Egyptian reliefs that this type of machinery was not used by Egyptian armies until the 13th century BC. In addition, it should be noted that these reliefs bore a clear relation to royal court circles, as did the literary descriptions of siege warfare in the royal inscriptions of Assyria and Egypt.[54] Accordingly, the same reservation noted above regarding the historical reliability of various details in the royal inscriptions equally applies to these reliefs. Moreover, like the royal inscriptions, the reliefs are fairly standardized; consequently, it is doubtful that they can provide specific information on any particular battle, e.g., the standardized depiction of most cities in the Assyrian reliefs, lacking any location-specific characteristics;[55] it is also doubtful whether all the military operations portrayed in a given relief were in fact used in the specific battle to which the relief was dedicated.[56]

53 In reconstructing the technological-military reality on the basis of the reliefs, one must take into account that the dimensions of various objects are not drawn to proportion, a well-known phenomenon in the art of the ancient Near East; thus, people on the reliefs appear approximately half the height of the wall, the relationship between the height of the wall and that of the ramming machine is also unrealistic and the soldiers in the ramming machine appear higher than the walls (cf., the depiction of the attack on Gezer in a relief of Tiglath-pileser III, in which an Assyrian soldier supposedly jumps from the ramming machine onto the wall).

54 On the ideological message and contents of the Neo-Assyrian reliefs and the parallels to the royal inscriptions, see Winter, *Studies in Visual Communication 7/2* (1981), 2-38; Reade, in *Power and Propaganda*, 329-343. The conclusions reached in these studies regarding the Assyrian reliefs are applicable to the Egyptian reliefs as well.

55 See Jacoby, *IEJ* 41 (1991), 112-131.

56 A rich assortment of means of warfare against a city is described in a single relief (e.g.,

IV. ARCHAEOLOGICAL EVIDENCE

There is archaeological evidence, in the form of layers of destruction, for the conquest of hundreds of cities, but few can be clearly connected to siege fighting. An archaeological find may be connected to a siege only if remnants are found in it that relate directly to the siege or the breakthrough stage.

The causeway laid against Tyre by Alexander the Great during the battle of 332 BC is one of the most striking pieces of evidence of siege in the ancient Near East. At that time, Tyre was an island, about 750 meters from the shore, a location which prevented any approach or attack using land combat techniques. The causeway took seven months to construct; it has become wider in the course of time, thanks to deposits of sand and gravel laid by the sea currents. Today, Tyre is a peninsula connected to the mainland by a land bridge some 500 meters wide.[57]

A siege ramp was found in the south-western corner of Tel Lachish, dating to the attack of Sennacherib, king of Assyria, on the city in 701 BC.[58] A counter ramp was also discovered on the same site, constructed by the defenders from inside the walls, in response to the approaching Assyrian ramp, as an additional obstacle for the attackers after the anticipated breach of the wall.[59] Only a few discoveries of this type, which can be clearly

---

Yadin, *Art of Warfare*, 346, 392–393, and compare the extended written descriptions, above, n. 3). In reliefs, the inclusion of so many different means of warfare – more than in written descriptions – may be due to the tendency to avoid empty spaces on the drawn surfaces.

57 For literary data concerning the building of the causeway by Alexander the Great and the tremendous difficulties encountered in this project, by natural forces and humans alike, see Diodorus 17.40–45; Curtius Rufus 4.2.8–3.15; Arrian 2.18–21.

58 This part of the rampart was excavated in the 1930s, but the excavators of Lachish at that period did not realize that it was a siege rampart. It was identified as such by Yadin in 1973, and was again excavated in 1977 and in 1983–85 (this time with a clear understanding of its nature) by D. Ussishkin. See Ussishkin, *The Conquest of Lachish by Sennacherib*, 49–55; and cf. in the following footnote.

59 For the final report concerning these ramparts, see: Ussishkin, *The Renewed Archaeological Excavations at Lachish (1973–1994)*, 695–767. On the dimensions of these ramparts and on the manner of their construction and purpose, see below, pp. 84–90.

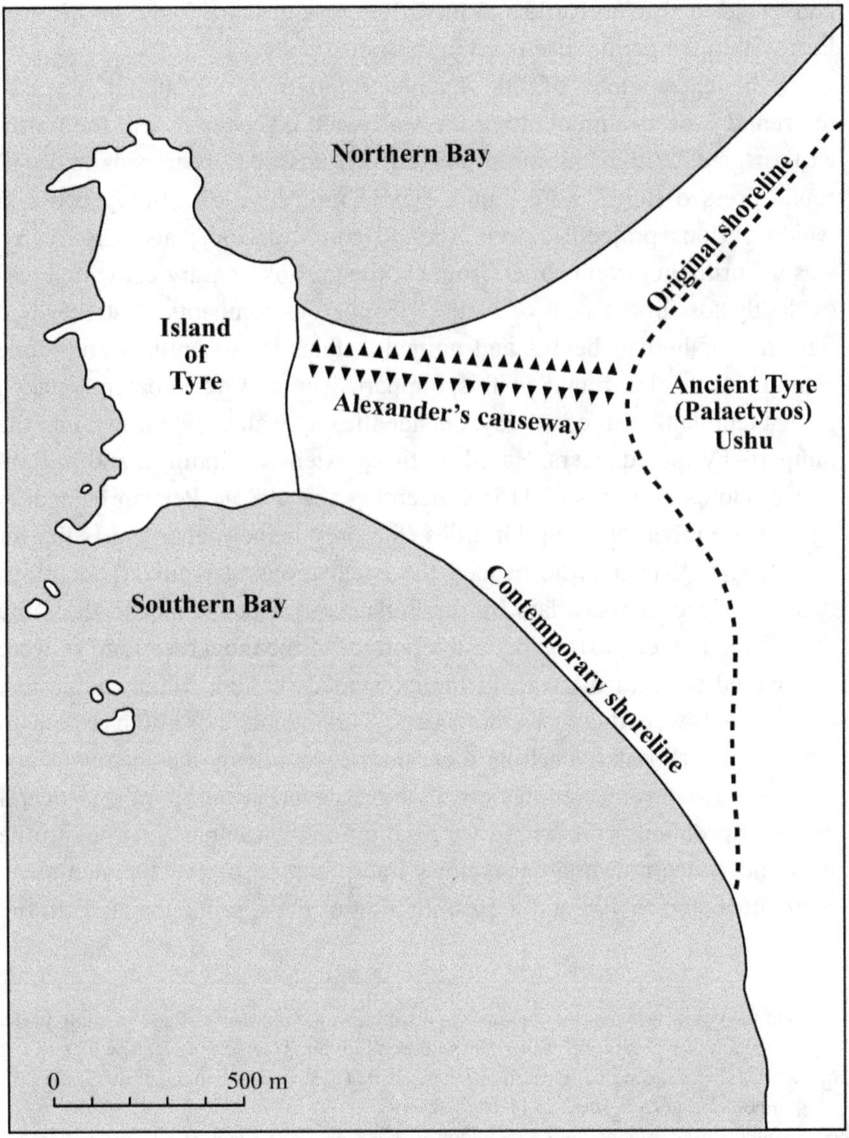

Fig. 2: Map of Tyre

connected to a given battle and its different stages, are found in the Near East within the period discussed in this book.[60]

In the excavations of Old Paphos (modern-day Kouklia), Cyprus, the remains of a ramp abutting the wall were discovered near the north-eastern gate. Within the ramp 449 arrowheads and spearheads and 442 round dressed stones were found, clearly indicative of a heavy battle in which various projectiles were shot to some distance; also discovered was a Corinthian iron helmet, from ca. the mid-6th century BC, which was evidently lost in the heat of battle.[61] Numerous remnants of statues and figurines of human beings and animals, steles, altars, pillars and stone capitals were also found in this rampart. These were apparently taken from a temple that stood nearby, outside the city walls, and thrown into the ramparts by the attackers. Based on these extensive findings, and in light of Herodotus' remarks (5.115) concerning the 498 BC Persian campaign against the rebellious Cypriot cities (the only name mentioned is that of Soloi), F. G. Maier (who headed the excavations at Kouklia) identifies this as a siege rampart, laid by the Persians against the city in the same war. Three tunnels, extending to the bottom of the moat (see Fig. 3), were discovered beneath the wall in the excavated section. Maier thinks that these tunnels were dug by the defenders, with the intent of filling them with flammable material and setting them on fire, to destroy the enemy's siege towers.[62] However, examination of the excavation report raises several essential problems concerning the relation of the rampart findings to the wall and concerning the excavator's interpretation of the "Persian siege ramp attributed to 498 BC."[63] Examination of these problems leads to the

60 On the conjecture that it is possible to distinguish the remnants of a siege ramp in the excavations of Mari (Tell Hariri) as well, see Salvini, *MARI* 5 (1987), 628–630.

61 On this helmet, see Snodgrass, in *Alt-Paphos auf Cypern: Ausgrabungen zur Geschichte von Stadt und Heiligtum, 1966–1984*, 43–49.

62 Maier and Wartburg, in *Archaeology in Cyprus, 1960–1985*, 141ff., esp. 152–159; Maier and Karageorghis, *Paphos: History and Archaeology*, 186–203.

63 There are a number of problems with Maier's approach: (a) The retaining wall bordering on the rampart touches the corner of the eastern tower (covering it, in fact), so it must be later and does not belong to the same period (Maier himself states that the retaining wall is late: from the middle of the 4th century BC). It follows, then, that the

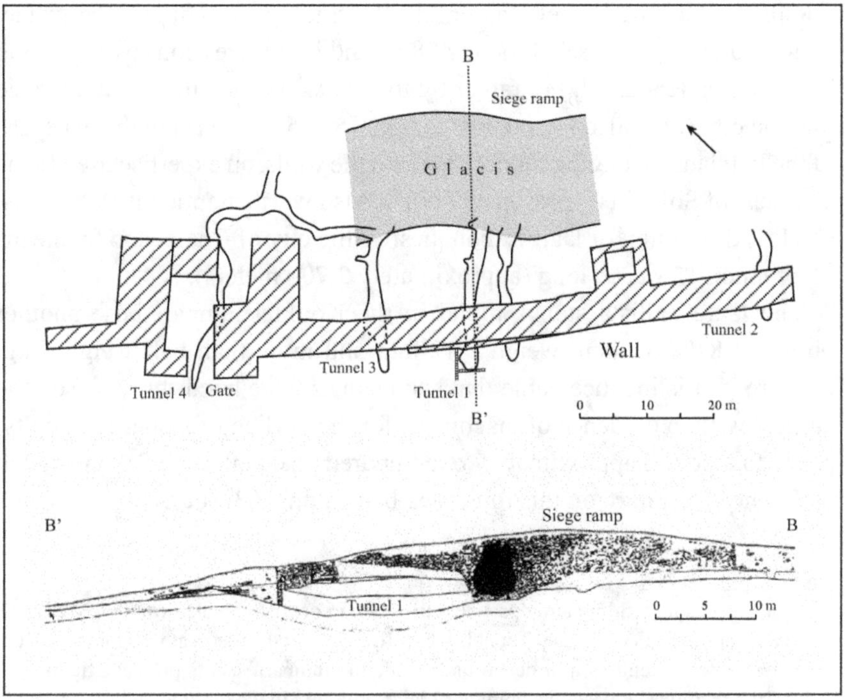

Fig. 3: Above: Diagram of the wall, the siege ramp and the tunnels in the gate area from the excavations of Kouklia (Old Paphos / Palaepaphos); below: cross-section B–B'

structure of the rampart as it appears today is only part of the original one. If Paphos continued to exist after the battle in question, was the wall "breached in 498 BC" not repaired until the second half of the 4th century BC? And how is it that the remnants of the rampart were not removed when the wall was repaired? (b) The wall of the city only survived to a height of 2.30 meters, and its original height is unknown. From the cross-section of the rampart, one can reconstruct the manner in which the earth and stones were piled up in the mound that was excavated, but it is impossible to determine the height of the rampart, its shape or the angle of its incline. It makes sense that its wide base was at the point farthest from the wall, that is, opposite to its present form; compare the siege rampart at Lachish (see above, nn. 58, 59). (c) The location of the rampart next to the gate seems surprising, as it draws fire from one of its most fortified bastions (Bastion I, which protrudes 14 meters from the line of the wall) and from another bastion, about 43 meters distant from it (did topographical conditions require this configuration?). According to Maier, the Persians broke through the gate (but is

possibility that the tunnels (assuming the findings at Kouklia are in fact related to the Cypriot rebellion in 498 BC and its suppression) were not dug by the defenders at all, but rather by the attackers, as a means of advance into the city (note also Herodotus' remark [5.115] concerning the expertise of the Persians in digging tunnels beneath the walls, an expertise used in the conquest of Soloi in Cyprus in 498 BC).[64] Also worth mention is Polyaenus' (7.11.5) discussion of Darius' conquest of the city of Chalcedon by means of a tunnel, 15 *stadia* long (approximately 2,700 meters).

The large ballista stones found in the Kouklia rampart pose another problem: 84% of them weigh 2–12 kg., and more than half weigh 6 kg. or more.[65] It is inconceivable that such stones were hand-thrown. Yet, the earliest written evidence of machinery for the hurling of stones (catapults) refers to a period approximately one hundred years after the Persian attack on Paphos (for more on this topic, see below, pp. 103–106).

---

there sense in building a rampart next to a gate in order to break through it?). (d) The excavator's conjecture, that Tunnels I and III, discovered underneath the rampart, were dug by the defenders in order to undermine the rampart, seems problematic in light of the relations between the amount of labor invested in the digging and the projected damage to the rampart (in fact, there is no evidence that fire was lit inside the tunnels). The two tunnels connect outside the wall beside the moat. It is reasonable to assume that this connection was of importance at the stage of digging, in order to advance from outside into the city rather than in the opposite direction. Moreover, if the tunnels were defense tunnels, what was the purpose of Tunnel II, located more than 35 meters to the east of Tunnel I, beyond the edge of the uncovered rampart? The assumption that this was also a defense tunnel implies that the width of the siege rampart was far greater than the extant remnants (and that most of the rampart was removed following the battle, and therefore not preserved), or that the tunnel was intended for sorties from within the city (if we accept Maier's statement that the tunnels were dug from within the city outwards, and not from the outside into the city), and not in order to undermine the ramparts.

64 On the other hand, the fact that the tunnels end very close to the wall inside the city may support the assumption that they were dug by the besieged, who did not dig more than necessary. Had these tunnels been dug by the attackers, it seems likely that they would have penetrated more deeply into the city, and not remained so close to the wall and the gate.

65 For details of the number of stones and their weights, as well as for theories regarding the manner in which they were hurled long distances, see Erdmann, *Nordosttor und persische Belagerungsrampe in Alt-Paphos*, 80–82.

A survey of the archaeological evidence includes several dramatic finds of skeletons of humans who died during an assault on cities or shortly thereafter:

In excavations of three of the gates of Nineveh (the gates of Shamash, Adad and Halzi), evidence was found of preparations for an anticipated attack on the city in its final stage – the entrances to the gates were narrowed from seven to two meters, in order to make breaching into the city via the gates more difficult (in light of the great thickness of the walls, it seem clear that breaching the gate was preferable to breaching the walls). At the Halzi Gate, skeletons of more than a dozen people were discovered (including a 12-year-old youth, struck by an arrow in his leg, a three-year old child and a ten-month old baby) who were killed during the Median assault on Nineveh in 612 BC. They were found lying as they collapsed when injured during the fall of the city.[66] No breakthrough occurred at the other two gates.

Archaeological evidence suggestive of what the inhabitants of a conquered city could expect may be found in the Ashdod excavations, Area D, Stratum 3b.[67] Human bones were found at several loci on the site, which had been thrown into courtyards and buildings in the wake of a slaughter probably associated with the conquest of the city.[68] The most striking find, the remains of 2,434 individuals (538 of whom [22.1%] were aged fifteen and under) were uncovered in Locus 1151. In another locus, Number 1114, a mass grave was discovered containing the remains of 376 individuals (215 of whom [57.1%] were aged fifteen and under), some of whom showed signs of (possibly posthumous) amputation, together with the bones of animals. Skeletal remains revealing signs of beheading were discovered in other loci – Number 1115 (nine to ten individuals), 1113

66 For a preliminary publication of these findings, see Stronach, in *Assyria 1995*, 315–319; Pickworth, *Iraq* 67 (2005), 295–316.

67 On the findings in the excavations of Stratum 3b of Area D, see Bachi and Ben-Dov, *ʿAtiqot*, English Series 9–10 (1971), 88–105.

68 The skeletal remains were published by Haas, *ʿAtiqot*, English Series 9–10 (1971), 212–214. For photographs of the remnants of the scattered bones thrown in without any order, see *ibid.*, Plates no. XXXV, 3; XXXIX, 1-3.

(seven individuals), 1006 (forty-five individuals, one of which showed signs of beheading).[69]

Stratum 3b of Area D corresponds to general Stratum VIII of the excavations at Ashdod, whose destruction has been dated to the end of the 8th century BC. M. Dothan, the excavator at Ashdod, attributes the death of the slaughtered to Sargon's conquest of the city.[70]

In the 1930s excavations at Lachish, remains were found of more than 1,500 humans in four tombs (Numbers 107, 108, 116, and 120).[71] These remains were not interred in an orderly initial burial, but were apparently thrown in, having been collected in the aftermath of a mass death that befell the population of Lachish, which prevented orderly burial. 695 skulls and a small quantity of other bones were taken for anthropological examination.[72] Examination of these bones and of the tombs in which they were found yielded limited but important data, which should be considered in any discussion of the disaster that led to the death of the people of Lachish:[73]

1. The average age of the dead of Lachish whose remains have been examined is significantly lower than that found in other burial sites; consequently, it is likely that they did not die of natural circumstances.
2. At least 48.2% were women and children.
3. Only one skull showed signs of fatal injuries.
4. Some of the remains showed signs of burning (perhaps these people were in or near burning houses before their bodies were thrown into the caves).
5. The skulls found at Lachish differ from skulls found in other Iron Age sites in Judah, and are similar to those of people of Egyptian origin.[74]

69 It is interesting that signs of beheading were found in those loci in which the number of skeletons was relatively small (1006, 1113, 1115). It would seem that these people were put to death separately. One should note that in Locus 1113, the beheaded bodies were placed in two rows.

70 See Dothan, *ʿAtiqot*, English Series 9–10 (1971), 21.

71 A description of these caves appears in Tufnell, *Lachish* III, 187–196; cf. Ussishkin (below, n. 76).

72 For discussion of the finds and their measurements, see Risdon, *Biometrika* 31 (1939), 99–166.

73 On points 1 to 5, see Risdon, *ibid.* On points 6–7, see Tufnell (above, n. 71).

74 Thus, Risdon, *ibid.* This statement was contested by Keith, *PEQ* 72 (1940), 7–12,

6. The style of the burial caves in which the bones were found suggests that they were dug in the Late Bronze Age.
7. The ceramic remains found in the burial caves – particularly in Tomb 120, in which most of the skulls were found – are relatively sparse, and it is difficult to ascertain the chronological relation between them and the dead of Lachish.[75]

The excavators of Lachish tend to connect these finds to Sennacherib's battle against the city in 701 BC.[76] However, in light of (5) and (7) above, it is not impossible that these remains are to be attributed to the stage in which Level VII was destroyed, at the end of the Late Bronze Age, when the city was under Egyptian rule.

Regardless of the origin of the dead of Lachish, the skeletons in question provide concrete evidence for the population density in a besieged city. At the end of the 8th century BC, the area of Lachish was approximately 7.5 hectares, clearly defined by the city walls. Therefore, the average ratio of dead (whose remains were found in the four graves) per hectare is 200, close to the usual urban density index in Palestine and in other areas of the ancient Near East during the period in question.[77] In other words, the

according to whom the skeletons found in Lachish are of the Mediterranean type (which also includes the inhabitants of Egypt, at least those of Lower Egypt). To confirm the statement that the dead of Lachish included people of an Egyptian-Nubian origin, but without discussion of the relationship between them and the rest of the population in southern Palestine, see Keita, *American Journal of Physical Anthropology* 75 (1988), 375-390.

75 See Zimhoni, *Studies in the Iron Age Pottery of Israel*, 160-164.

76 Ussishkin, *The Conquest of Lachish by Sennacherib*, 56-58.

77 The population density of urban settlements in Palestine and in other areas of the Near East in ancient times is subject to scholarly debate. There are those who estimate it at approximately 400-500 people per hectare (see Shilo, *BASOR* 239 [1980], 25-35); by contrast, others estimate the density as only 200-250 per hectare (see Broshi and Finkelstein, *BASOR* 287 [1992], 47-48). As noted above, we tend toward the smaller estimate, particularly in light of the fact that 8th century BC Lachish had developed as a government center rather than as a crowded village settlement that grew, such as Tell en-Naṣbeh. The palace-fort complex in the center of the city, that occupied more than 1.3 hectare, is indicative of the planned nature of Lachish of that period as a government center.

number of dead in conquered Lachish was close to the overall number of inhabitants in a city of similar proportions.

If the skeletons found in the four tombs are indeed connected to Sennacherib's conquest of Lachish, two points should be emphasized: (1) estimates of the population density in besieged Lachish must take into account that not all those found within the city were killed and thrown into the four pits. Some inhabitants, men, women and children, must have gone into exile;[78] (2) about half of the above-mentioned bones found at Lachish were of civilians, not of combatants who died in battle. The absence of signs of fatal injuries supports the assumption that they died of famine or disease.[79]

78 See the sections of the relief on the battle of Lachish in Ussishkin, above, note 76 (for enhanced detail, see *ibid.*, figs. 68–70, also figs. 78, 80, 84–87). Compare also the epigraph for the description of the conquest of Lachish in the relief in question: "Sennacherib, King of the universe, King of Assyria, sat upon an armchair and the booty/captives (*šallatu*) of Lachish passed before him" (OIP II, 156).

79 The bones found at Lachish, if they are indeed the remains of people who perished en masse and were not brought to proper burial, illustrate the motif of bodies rolling about in the streets and the sight that greeted conquerors upon their entrance (cf. Isa. 51:20; Jer. 14:16; Lam. 2:11–12; 4:5; Streck, *Asb.* 38–40 iv 79–85; Brinkman, *JCS* 25 [1973], 93:7–9).

CHAPTER THREE

MILITARY ASPECTS

I. THE BLOCKADE

The imposition of a blockade on a city is intended to cut it off from supplies of food, water and war materiel (weapons, ammunition, fortification materials, medical supplies and the like), as well as to prevent the entrance of reinforcements and relief and the evacuation of the non-combatant population, whose presence in the city hampers its ability to withstand attack. The effect of a blockade is cumulative, so it must be continuous and generally extended in time (with the exception of cutting off water supply, which is rapid in impact; see below, pp. 64–66).[1]

The tightness of the blockade – the degree of constriction that the besieging forces can maintain – depends on the length of the front surrounding the city and on the attackers' ability and willingness to appropriate manpower and war materiel for the siege.[2] A hermetic siege requires forces to be positioned around the entire city, at maximum proximity to its external fortifications. The seizure of outposts at key points surrounding the city cannot, on its own, prevent the infiltration of individuals, but can prevent the movement of large groups and transports. A more economical (but less effective) method to limit movement near

1 For the attackers, blockade is preferable to breakthrough in the expected number of casualties; for discussion, see above, p. 1, n. 1. Yet, blockade entails other difficulties; for discussion, see below, pp. 106–110.

2 Thus, for example, the circumference of a circular-shaped city whose area is 65 hectares (e.g., the area of Jerusalem during Nebuchadnezzar's siege) would be 2,860 meters, whereas a square-shaped city of a similar area would have a circumference of 3,224 meters. The length of the walls of Nineveh was 12 km.

the city is to use mobile forces or interspersed roadblocks to block the movement of army troops and supply loads at key points some distance from the city.

Egyptian royal inscriptions contain descriptions of hermetic blockades, reminiscent of the type of blockade used by the Roman army.[3] Details on the operation of this method of closure appear in the following description of the seven-month siege imposed by Thutmosis III on Megiddo:

> [Orders were given to] the troop-commanders to [provide for their soldiers and to let] every [man know] his place. They measured the town, surrounded (it) with a ditch,[4] and walled (it) up with fresh timber from all their fruit trees.[5] His majesty himself was on the fort east of the town, guarding [it day and night]… [surrounded] by a thick wall… It was given the name "Menkheperre-Encircler-of-Asiatics." Guards were placed at the camp of his majesty and were told: "Steadfast, steadfast! Vigilant, vigilant!"… His majesty… [No one] of them [was permitted to come] out beyond this wall, except for a coming out to knock at the gate of their fortress (in order to surrender).[6]

Tefnakht's siege on Hnes (Heracleopolis Magna), nearly 800 years later, is described in almost identical manner in the Pi(ʿankh)y Stele:

> Now [he is] besieging Hnes. He has encircled it completely (lit., "he made himself into a tail-in-the-mouth"; i.e., he encircled the town like a coiled snake), not letting goers go, not letting entrants enter, and fighting every day. He has measured it in its whole circuit. Every

3 Evidence of the hermetic blockades imposed by the Roman army are found in writing (see, e.g., Caesar, *Gallic Wars*, 7.72–74; Appian 5.33; 6.90) and in archaeological remains of siege camps (found, for example, in Masada and in Numantia in Spain).

4 Compare also: "The soldiers set up a camp, dig a channel (*ḫirītu*), surround the city in a camp" (ARM I 90:18–21).

5 It should be noted that of all fruit trees that grow in Palestine, only the date palm can be used for construction. Hence, it seems that the wall referred to in connection with the siege of Megiddo was no more than a barricade into which tree branches were inserted, rather than trunks and beams. See discussion below, p. 39, note 15.

6 Lichtheim, *Ancient Egyptian Literature*. Vol. II: *The New Kingdom*, 33.

> count knows his wall (i.e., each chief allied with him is camped before a section of the wall). He has made every man besiege his portion, to wit the counts and rulers of domains (ll. 4–5).[7]

The nature of this hermetic siege is reflected in an order sent by Pi(ᶜankh)y to his officers:

> Enter combat, engage in battle; surround... capture its people, its cattle, its ships on the river! Let not the farmers go to the field, let not the plowmen plow. Beset the Hare nome; fight against it daily! (ll. 8–9).[8]

The method of isolating a city by seizing outposts at critical points around it is mentioned in Assyrian royal inscriptions. In the description of the conquest of the cities of Judah, through various methods of assault, the Sennacherib inscriptions state:

> Himself (Hezekiah, king of Judah) I made a prisoner in Jerusalem, his royal residence, like a bird in a cage.[9] I blocked him with fortified posts and made departure via the gate of his city into an unbearable ordeal.[10]

The same method appears in descriptions of Esarhaddon's "tenth campaign" and Ashurbanipal's "third campaign," both directed against Baal, king of Tyre (see below, p. 41).

7 Lichtheim, *ibid.*, Vol. III*: The Late Period*, 68.

8 *Idem.*, *ibid.*, III., 69.

9 The image to confine "like a bird in a cage" (*kīma iṣṣūr quppi esēru*) also appears in the annals of Tiglath-pileser III, relating to the Assyrian siege on Rezin, king of Damascus (Tadmor, *ITP* 78:11). On the conceptual background for the use of this expression in the Assyrian royal inscriptions, see Tadmor, *Zion* 50 (1985), 74–75. A similar image appears in the letters of Rib-Haddi, ruler of Byblos, to Pharaoh, in which he reports on the land and sea blockade imposed on Byblos and its environs, complaining that he, his *ḫu[pšu]* people, and the city of Ṣumur (that was within his realm), are like "a bird in a cage": *kīma iṣṣūri ša ina ḫuḫāri/ kilūbi šaknat*, EA 74:45–48; 79:35–38; 81:33–36; 105:8–10; 116:17–20.

10 OIP II 33 iii 27–30. On the reading ḪAL.ṢU.MEŠ = *birāti*, see Borger, *BiOr* 32 (1975), 71.

In the description of the Assyrian siege on the city of Nisibis (896 BC), the annals of Adad-nirari II mention two methods of blockade – short-range and intermediate-range:

> I confined Nur-Adad the Temannu in the city of Nisibis (and) established seven redoubts (*ālāni*) around it[11]... I encircled his moat with my warriors like a flame... (and) deprived him of grain.[12]

The Hebrew term דָּיֵק, the meaning of which is debated by scholars,[13] is apparently connected to siege warfare. In the Bible, the root בנה (to build/ construct) is used for the construction of a דיק, whereas the construction of a siege ramp is described with the verb שפך (cast up; cf. Ezek. 4:2; 17:17; 21:27). In 2 Kings 25:1//Jeremiah 52:4, it is said: "... Nebuchadnezzar moved against Jerusalem with his whole army. He besieged it, and they built a דיק (a siege-wall?) against it all around." This formulation (as well as the Aramaic translation of דיק in the Bible as כרכום [in Peshitta to Jer. 52:4; Ezek. 4:2; כלכומא]) brings the Semitic term closer to the Latin/English "circumvallation," even if its exact nature during the period and region discussed in this book is not entirely clear. The Akkadian word *dāiqu*, parallel to Hebrew דיק, has been found only once thus far, in Esarhaddon's "Letter to the God," describing his war with Ik-Teshub, king of Shubria and the Assyrian siege on the city Uppume. The text details a disastrous attempt by the defenders to set fire to the Assyrian siege ramp, which presumably contained wood (an attempt which led, instead, to the burning down of the city wall due to a change in wind direction), and continues: "[My troops?] crossed over the *dāiqu* in order to give battle… attacked and defeated them (the defenders)."[14] The *dāiqu*, which the Assyrian soldiers crossed during the active stage of battle, seems to have been an obstacle or

11 For this meaning, see CAD A/1 387 *ālu* 4.

12 RIMA 2 151:63, 66, 68.

13 Driver suggests that the term דיק refers to a portable blockade device ("an engine of attack into which men climbed"), while כר is the battering ram; see Driver, *Biblica* 35 (1954), 147–148. This suggestion does not correspond to Esarhaddon's description of his war against Uppume; see the next note.

14 Borger, *Asarh.* 104:8–9 *[…] epēš qabli u tāḫāzi dāiqu ibbalkitūni […]-iš itbûšunuma iškunū taḫtâšun.*

parting-wall separating them from the city wall, which they assaulted by climbing. Hence, the term דיק ought to be interpreted as a structure built by the besiegers in order to close in on the circumference of the city, as in the above quotations from the inscriptions of Thutmosis III and Pi(ᶜankh)y.[15]

The isolation of besieged cities from their distant surroundings by the use of chariots is reflected in sources from the Hittite milieu. In the story of the siege of Urshu, the Hittite king orders the attackers to guard the roads of the besieged city to prevent anyone from entering or leaving, and is told that eighty chariots and eight armies surround the city (rev. 23–27).[16] Similarly, the author of the *General's Letter* writes to his king that for five months he has been guarding the roads and passages to the land of Amurru night and day with chariots, half of which are deployed in the coastal area and the other half in the foothills of the Lebanon. One of his tasks is to prevent Egyptian provision of foods (*ukullû*) and soldiers (*ṣābu*), from reaching the city of Ardat.[17]

Another method of imposing a blockade from afar is described in the annals of Sargon II, concerning his war in southeastern Babylonia (710 BC):

> I dammed the Tupliash River, the river of their confidence, with heaps of earth and reeds. I constructed two fortresses (*birāti*), one opposite the other, and I caused distress to come over them. They came out of the midst of the Uknu River and seized my feet.[18]

This siege was directed against Aramean tribes in Babylonia, at least some of whom were not sedentary; however, there is no doubt that a siege of this

15 The verb בנה (to construct) and the reference to the use of trees – in connection with the activity of the attackers – also occur in Deut. 20:20: "Only trees which you know do not yield food may be destroyed. You may cut them down *for constructing siegeworks* against the city (ובנית מצור על העיר) that is waging war on you..." It seems that here too the reference is to a hermetic blockade, as stated above.

16 Güterbock, *ZA* 44 (1938), 122–124; Beckman, *JCS* 47 (1995), 25.

17 For a transcription and translation of the letter, see: Izreʾel, in Izre'el and Singer, *The General's Letter from Ugarit*, 22–27. On the geography of the region which the author of the letter guarded, see: Singer, *ibid.*, 117–121. On the date of the letter and its political and military background, see Singer, *ibid.*, 170–183 (and references therein).

18 Lie, *Sar.* 48:3–4.

type could also be applied against many other cities located on rivers or canals, particularly in Mesopotamia and in Egypt.[19]

Another form of blockade is unique to coastal cities and islands. The restrictions imposed by the lay of the land do not apply to sailing; therefore, marine transport is much more flexible and less subject to blockades than land transport. This is an advantage of islands or coastal cities in the defense against blockade. The ability of ships to break through a naval blockade depends to a large extent on their seamanship and the technical qualities of their vessels; unlike in land warfare, the size of the troops is of lesser importance.[20] Most military organizations in the ancient Near East did not have their own naval forces (unlike those of the Aegean region); consequently, their ability to impose a naval siege was limited, even non-existent in practice, unless they had other naval fleets available (Phoenician, Cypriot, Egyptian and, of course, Ionian).

The difficulties encountered by land forces in operating a naval blockade are revealed in the writings discussing the blockades imposed on Tyre. The city of Tyre was located on a rocky island, some 750 meters from the mainland. Water was brought to the island in boats from the nearby coast.[21] According to Menander of Ephesus, quoted in Josephus' *Antiquities* 9.14.2, Shalmaneser V of Assyria (727–722 BC) besieged Tyre and "placed guards at the river and the aqueducts (on the sea shore opposite the island) to prevent the Tyrians from drawing water; they endured this for five years, and drank from wells which they had dug." It is hard to imagine that the inhabitants of the island were able to withstand such a prolonged siege without supplies of food, fuel, and other essential materials.[22] It appears

19 For an archaeological find that may be related to this feature, compare: Young, *Bulletin of the Society for Mesopotamian Studies* 6 (1983), 25, 28–30.

20 For example, *Antiquities* 9.14.2 relates that when Shalmaneser V fought against Tyre, the Tyrians, with their 12 ships, defeated the 60 ships that the Phoenicians had made available to the Assyrians for the siege.

21 See Papyrus Anastasi I xxi 1–2: "They say another town is in the sea, named Tyre-the-Port. Water is taken (to) it by boats, and it is richer in fish than sands" (cf. *ANET* 477).

22 On the need to bring water, wood and straw to Tyre, that was subjected to a blockade after it was cut off from Ushu (Palaetyros, Old Tyre, that was opposite the island), see

that Tyrian naval superiority enabled the Tyrians to provide themselves with supplies from more distant sources. A similar situation is depicted in the inscriptions of Kings Esarhaddon and Ashurbanipal of Assyria. The account of Esarhaddon's "tenth campaign," against Baal, king of Tyre, states: "I blocked him by fortified posts (*birāti*) and cut off food and water with which they sustain their life."[23] The geographical location of Tyre is reflected in the description of Ashurbanipal's "third campaign," also directed against Baal, king of Tyre: "I blocked him by fortified posts and seized his approaches by sea and land. I constricted their throats and subjected them to my yoke."[24]

Citing Philostratus (see also *Against Apion* 1.21), Josephus writes (*Antiquities* 10.11.1) that Nebuchadnezzar laid siege to Tyre for thirteen years. Scholars usually relate the following passage to this incident:

> In the twenty-seventh year, on the first day of the first month, the word of the Lord came to me: Son of Man, King Nebuchadrezzar of Babylon has made his army expend vast labor on Tyre; every head is rubbed bald and every shoulder scraped. But he and his army have had no return for the labor he expended on Tyre (Ezek. 29:17–18).

We have no evidence of Nebuchadnezzar's naval activity against the island of Tyre, and on the face of it, this text can be perceived as invoking an image of land warfare. The prophet ignores the technical and military data characteristic of the island of Tyre; similarly, in Ezek. 26:7–14, explicit

---

the letters of Abi-Milku of Tyre to Pharaoh (EA 147:61–66; 148:30–34; 149:49–53; 151:37–44; 154:11–18; 155:17–23). Abi-Milku also complains that there is no place in Tyre to bury the dead, and that the inhabitants of the island are prevented from burying their dead in Ushu (EA 149:52–53; 155:20–21). The inability to bury people in their ancestral graves, located outside the inhabited area, and the need to bury them inside the besieged city, certainly had far-reaching emotional consequences and possibly also deep spiritual-religious implications. On the assumption that these documents detailing the shortage of basic supplies do not necessarily reflect the situation within Tyre, and may be a reflection of standard hardship formulae commonly used in Canaanite literature, see: Gevirtz, *Orientalia* 42 (1973), 165–167.

23 Borger, *Asarh.* 112:14.

24 Streck, *Asb.* 16 ii 52–55.

components of land warfare are incorporated with geographical elements typical of Tyre:

> For thus said the Lord God: I will bring from the north, against Tyre, King Nebuchadrezzar of Babylon, a king of kings, with horses, chariots, and horsemen – a great mass of troops. Your daughter-towns in the open country he shall put to the sword; he shall erect a דיק (circumvallation/palisade?) against you, and shall cast up siege-ramps against you and shall raise a screen of shields (צנה) against you. He shall turn the force of his battering rams against your walls and smash your towers with his swords. From the cloud raised by his horses dust shall cover you; from the clatter of horsemen and wheels and chariots, your walls will quake when he enters your gates as men enter a breached city. With the hoofs of his steeds he shall trample all your streets…

Tyre was finally conquered by Alexander the Great (in 332 BC) at the end of a siege that lasted for seven months, during which Alexander built a causeway from the shore to the island, using typical methods of land warfare, after being joined by the navies of Cyprus and the other Phoenician cities.[25]

The uniqueness of coastal cities lies in the equal availability of both land and sea transportation. This forces attackers to isolate the coastal cities on land and sea alike (whereas the blockade of an island requires only blocking sea transport). Thus, for example, Rib-Haddi of Byblos complains in his letter to Pharaoh that a blockade was imposed on the city of Ṣumur (that was within his realm) by the sons of Abdi-Ashirta, ruler of Amurru, on dry land, and by the people of Arvad from the sea (*EA* 105:11–13). Similarly, the Persians laid siege to Miletus both on land and on the sea (494 BC; Herodotus 6.18), as did Alexander against that same city (333 BC; Diodorus 17.22.3). On more than one occasion, the besiegers only had the means

25 It should be noted that the works concerning Alexander's warfare against Tyre make no mention of hardship – shortage of water or food – in the city. Alexander did not defeat Tyre by means of a blockade, but by techniques of breakthrough; see Diodorus 17.40–46; Arrian, 2.18–24; Curtius Rufus 4.2.18–4.14.

for one type of warfare available to them. In such cases, the blockade was not effective, and the besieged cities generally held their own, even for prolonged periods.[26]

1. Negotiation

Negotiation, conducted throughout most stages of the siege, was an integral component of the contact between the two parties in siege warfare. At the beginning of the siege, negotiation was undertaken in an attempt to arrive at an agreement without resorting to fighting, which would entail casualties, effort and damage to both sides.[27] During the siege itself, negotiations

26 On Tyre's sustained resistance to the sieges by Shalmaneser V and Nebuchadnezzar, see below, n. 228. An example of the use of sailing vessels to conquer a city may be found in the attack of Pi(ʿankh)y on Memphis, which was located on the Nile and defended by an enormous wall. Rather than building a rampart against it and capturing it by extended and difficult land combat, he preferred to advance his warriors on the Nile and attack in the port area (which was not defended by land obstacles, unlike other parts of the city). See Pi(ʿankh)y's Inscription, ll. 94–96; and Darnell, in *Ägypten im afro-orientalischen Kontext: Aufsätze zurArchäologie, Geschichte und Sprache eines unbegrenzten Raumes*, 73–84.

27 For examples of offers made by the attacker to the besieged, in the beginning of the siege, to surrender and thereby save themselves from a greater disaster, see below, pp. 44–48. At times, the defenders, believing that they would be delivered soon, asked that hostile activity against them be delayed for a certain period, declaring that they would surrender without struggle if help did not arrive by the end of the designated period. On more than one occasion, the attackers agreed to such a suggestion, confident that no help would arrive and wishing to avoid their own casualties. See 1 Samuel 11:1–3 for the suggestion of the people of Jabesh-gilead to Nahash the Ammonite, that he should refrain from activity against them for seven days while they send messengers "throughout the territory of Israel" (a suggestion which was accepted). The same motif appears in the Athena Temple Chronicle of Lindos on the Isle of Rhodes: the inhabitants of besieged Lindos, whose water supplies were running out, proposed a five-day ceasefire to Datis, commander of the navy that King Darius of Persia had sent against them and other rebellious Ionian cities, promising that if they were not delivered within this period of time then they would surrender. Datis agreed, but the city was saved on the following day when heavy rain fell, providing an abundance of water for the besieged (for the Greek text of this section of the chronicle, see Blinkenberg, *Lindos, Fouilles de l'Acropole 1902–1914, II. Inscriptions,* vol. I 181–184; for an English translation, see Heltzer, *La Parola del Passato* 44 [1989], 90–91). According to Mari letter A.319, the elders of the city of Razama went out to King Atamrum of Allahad ten days after the latter besieged their city, declaring their

were generally conducted for humanitarian reasons (such as a ceasefire to evacuate the deceased from areas under enemy fire);[28] toward the end of the siege, negotiation was intended to obtain favorable conditions, either in light of the city's diminishing chances to avoid conquest or, once it became clear to the attacker that he could not defeat the city successfully by force, he preferred partial victory to total failure.[29]

Negotiations could be conducted in mutual respect and even congeniality or through intimidation, depending on the prevailing military and political conditions.[30]

(a) "Friendly Words"[31]

An attempt to persuade a city to surrender by declaring one's good intentions and assuring protection for its inhabitants, while refraining from any threat against them, appears in the appeal made by Pi(ᶜankh)y, through

---

willingness to make peace (i.e., to surrender) if he would remove his army to a distance of one hour from the city (ll. 11–14). From the sequel of the letter, it appears that the besieged anticipated the arrival of King Zimri-Lim of Mari and his army to save them within a matter of days, and that the negotiations were only intended to buy time until the arrival of the liberating army (ll. 38ff.); cf. Charpin, *MARI* 7 (1993), 197–203.

28 For example, Alexander the Great's siege on Halicarnassus (Diodorus 17.25.6).

29 Thus, for example, Hezekiah's conditions of surrender to Sennacherib, that were surely determined before his army left Judah and established the amounts of tribute, enumerated the people sent by the king of Judah to Assyria and may have also outlined the boundaries of the area which was taken from the kingdom of Judah and annexed to its neighbors in Philistia. It is, nevertheless, difficult to determine the exact circumstances of these negotiations, since the question whether the Assyrian army did in fact subject Jerusalem to a real siege remains unanswered. Likewise, the conditions of King Jehoiachin's surrender to Nebuchadnezzar, the list of people who went into exile with him (see 2 Kgs. 24:10ff.) and his future status in Babylonia (known from administrative documents discovered in the excavations of Nebuchadnezzar's palace), all these were probably determined before he surrendered to the king of Babylonia.

30 On the preference for negotiations, whether amicable or through intimidation over conquest by force, see e.g., Prov. 21:22: "One wise man prevailed over a city of warriors and brought down its mighty stronghold."

31 For the Akkadian forms of this term, compare *ina pî ṭābi*; *ina dibbī ṭābūti*; *ina salīm ṭubbāti*: above, p. 21, n. 28. On the view that "kind words" are not simply "friendly words," but rather an official act of granting rights to a city (or public), see Weinfeld (in wake of W. L. Moran), in *Maarav* 3 (1982), 27–53.

messengers, to Memphis during his war against Tefnakht, ruler of Northern Egypt:

> Do not close (your gates), do not fight, O home of Shu since the beginning! Let the entrant enter, the goer go; those who would leave shall not be hindered! I shall offer an oblation to Ptah and the gods of Memphis. I shall sacrifice to Sokar in Shetit. I shall see South-of-his-Wall (=Ptah), and I shall sail north in peace!... (The people of) Memphis will be safe and sound; one will not weep over children. Look to the nomes of the South! No one was slain there, except the rebels who had blasphemed god; the traitors were executed (ll. 85-86).[32]

In similar fashion, when Babylon under the leadership of Ukin-zer the Chaldean closed its gates to the messengers of Tiglath-pileser III, the messengers promised the inhabitants of the city that the king of Assyria would come to their city and confirm their right of *kidinnūtu* (a preferred status granted to several Babylonian and Assyrian cities, by which the inhabitants enjoyed protection from bodily harm and exemption from various taxes).[33]

A declaration of good intention on the part of the aggressor may be heard in the words of David's general, Joab son of Zeruiah, to the wise woman from the besieged town of Abel-beth-maacah. In answer to her question, "Do you seek to destroy a mother city in Israel? Why will you swallow up the Lord's possession?" he replied, "Far be it from me that I should swallow up or destroy. Not at all. But a certain man from the hill-country

32 Grimal, *La stèle triomphale de Pi(ʿankh)y au Musée du Caire,* 94; Lichtheim, *Ancient Egyptian Literature* III, 75.

33 According to NL 1 (=CTN V pp. 19-20), Tiglath-pileser's messengers spoke to "the citizens of Babylon" (*mārē Babili*) who stood in the Marduk Gate of Babylon alongside "Asinu the servant of Ukin-zer" the Chaldean (ll. 5-12; the promise of the *kidinnūtu* appears in line 17). It therefore seems that the representatives of the king of Assyria attempted to make the "Babylonians" rebel, in a manner similar to that intended by Rabshakeh in Jerusalem. See below, pp. 54-57. For the first publication of this letter, see: Saggs, *Iraq* 17 (1955), 23-26; for corrections and improvements in the reading, see Saggs, in *Archaeology and The Old Testament Study,* 47; idem, *Assyriology and the Study of The Old Testament*, 17-18; von Soden, *Bibel und Alter Orient,* 152-154.

of Ephraim, Sheba son of Bichri, has rebelled against King David; give up him alone, and I will withdraw from the city" (2 Sam. 20:19–21).

Persuasion was required not only of the aggressors; sometimes, a city was saved from conquest by the wits of its representatives in negotiation with the attackers. The story of Sheba son of Bichri's rebellion includes a lively depiction of successful negotiations conducted between Joab and the wise woman, at the latter's initiative. After hearing Joab's conditions, the woman "in her wisdom approached all the people," and convinced them to put Sheba to death and throw his head to Joab. "So he sounded the ram's horn, and they dispersed from the city, every man to his tent" (*ibid.*, vv. 16–22).[34]

The terms used, "friendly words" (*dibbī ṭabūti*), "peaceful negotiation" (*salīm ṭubbāti*) and the assurance of peace to the inhabitants, are misleading. At first glance, these terms seem to suggest goodwill[35] and reciprocity between the negotiating parties; yet the true meaning behind the words usually refers to the surrender of the city. The true meaning of the peace expected from such negotiations is reflected in Deuteronomy 20:10–11:

> When you approach a town to attack it, you shall offer it terms of peace. If it responds peaceably and lets you in, all the people present there shall serve you at forced labor.[36]

Thus, the elders of Razama tell King Atamrum of Allahad who besieged their city, "We are for peace," and offer him money for moving his camp one hour's distance from the city (A.319:11–14).[37] A similar situation is implied in Sennacherib's offer, as articulated by Rabshakeh, who addresses "the people who are sitting on the wall" of Jerusalem:

34 The statement "wisdom is better than valor" suggests that the passage in Eccl. 9:14–15 on the poor man whose wisdom saved the city from a siege by a great king, also refers to an achievement through negotiations.

35 Compare Deut. 2:26ff.: "Then I sent messengers from the wilderness of Kedemoth to Sihon, king of Heshbon, with these peaceful words: 'Let me pass through your country. I will keep strictly to the highway, turning off neither to the right nor to the left…'"

36 For discussion of calling on a city for peace after it has been surrounded, see ARM II 42:8 (above, p. 15, n. 16).

37 Charpin (above, n. 27), 199.

> Make your peace with me and come out to me, so that you may all eat from your vines and your fig trees and drink water from your cisterns, until I come (at the end of the stage of combat of that same campaign) and take you away to a land like your own, a land of grain and new wine, of bread and vineyards, of olive oil and honey (2 Kgs. 18:31–32 // Isa. 36:15–16).

The true meaning of "making peace" in this context is none other than exile, with all its implications.

Consequently, the terms "peace" (in Deut. 20:10–11 and in the Assyrian queries to Shamash) and "making peace" (in 2 Kgs. 18:31–32) seem to be euphemisms for subjugation, as implied by the very circumstances in which these negotiations were carried out. However, it is important to recall that the Semitic term "peace" (שלום, *šulmu*) means only "lack of hostilities" or "non-aggression" between political entities, without defining the exact balance of power or political relationship between them (similarly, the term ברית "covenant" is used in the Bible simply in reference to a contract, whether the parties are equals or not). One must also be aware that subjugation, and even exile, as proposed in the context discussed here, are preferable to the alternative of the city's conquest following a refusal to surrender:

> If it does not surrender to you, but would join battle with you, you shall lay siege to it. And when the Lord your God delivers it into your hand, you shall put all its males to the sword. You may, however, take as your booty the women, the children, the livestock, and everything in the town, all its spoil; and enjoy the use of the spoil of your enemy which the Lord your God gives you (Deut. 20:12–14). [38]

Rabshakeh's comments at the end of his appeal to the besieged people, "Stay alive and do not die" (2 Kgs. 18:32), should be understood in the

38 This is also seen in Josh. 9:15: "Joshua established peace with them, he made a pact with them to spare their lives." Turning the Gibeonites to "hewers of wood and drawers of water to the entire congregation," after their scheme had been discovered, did not involve any violation of the covenant or the oath of peace that had been made with them.

same spirit.[39] A similar structure is evident in the reply of the leaders of Melos to the representatives of Athens who came to negotiate Melos' surrender (416 BC):

> The outcome of the discussion will in all likelihood be, if we win the debate by the righteousness of our cause and for that very reason refuse to yield, war for us, whereas if we are persuaded, servitude (Thucydides, 5. 86).

The awareness that saving life in itself (even with potential exile) is preferable to death by sword, famine, or plague in the besieged city, is expressed in Jeremiah 21: 8–9; 38:2.

(b) Threats and Intimidation

The attacker addresses the besieged city with a demonstration of force to emphasize his determination and the defenders' inability to withstand it. Thus, Sennacherib sends his representatives – the Tartan, the Rabsaris and the Rabshakeh – "from Lachish with a large force to King Hezekiah in Jerusalem" (2 Kgs. 18:17; cf. Isa. 36:2); Rabshakeh calls their attention to the striking difference in the sizes of the Assyrian and Judean armies, and argues that help from Egypt is unlikely:

> Thus said the Great King, the King of Assyria: What makes you so confident? You must think that mere talk is counsel and valor for war! Look, on whom are you relying, that you have rebelled against me? You rely, of all things, on Egypt, that splintered reed of a staff, which enters and punctures the palm of anyone who leans on it! That's what Pharaoh king of Egypt is like to all who rely on him… Come now, make this wager with my master, the king of Assyria: I'll give you two thousand horses if you can produce riders to mount them. So how could you refuse anything, even to the deputy of the least of one

39 The words "stay alive and do not die" do not appear in the parallel version in Isa. 36:16–17. It would seem that this claim in the Book of Kings is an addition, intended to explain the attractiveness of Rabshakeh's offer (surrender and exile), which does not seem appealing on the face of it.

> of my master's lesser servants, relying upon Egypt for chariots and horsemen?" (2 Kgs. 18:19–21, 23–24).[40]

The Chronicler, in his version of Sennacherib's war against Judah, describes the comments of the "servants of Sennacherib" (Rabshakeh himself is not mentioned in this book) as follows: "They called loudly in the language of Judah (i.e., Hebrew) to the people of Jerusalem who were on the wall, to frighten them into panic, so as to capture the city" (2 Chr. 32:18).

Similar things were said by the representatives of Athens to the leaders of Melos: (a) the weak have no chance of standing up against those who are far stronger than them (Thucydides 5.101); (b) the hope for success in war without substantial force can only bring disaster upon those who place all their hopes upon it (*ibid*., 102–103); (c) one cannot rely upon the Spartans to come to the aid of Melos, although the people of Melos were Spartan in origin and their allies, as Spartans prefer their own advantage and security above justice and honor (*ibid*., 104–110). At the end of these negotiations, the Athenians say to the leaders of Melos:

> When a choice is given you of war or of safety, do not hold out stubbornly for the worse alternative... Consider, then, once more

40 Rabshakeh's substantive arguments, whether "friendly words" or intimidation, have parallels in various episodes in the history of the ancient Near East and Greece. By contrast, the theological arguments found in 2 Kgs. 18: 22, 25, 30, 32b-35 and 19:10–13 (= Isa. 36:7, 10, 15, 18–20; 37:10–13) differ from the typical convincing arguments known from ancient history, and their relation to military and political reality is problematic (cf. above, p. 12, n. 10). Parallels to these may be found in the Assyrian ideology, in works intended to explain *post factum* the grave actions taken by the Assyrian kings against Babylonia. One such example is the anger of the gods of Babylon against its inhabitants during the reign of the Kassite king Kashtiliash, which led the gods to abandon their cities (leaving them undefended) and to support King Tukulti-Ninurta I of Assyria, who fought against Kashtiliash, under the gods' orders; see the *Tukulti-Ninurta Epic*, Machinist edition: i B 32'–47', A 10'–27'; iv 40'–51'; v 31'-40' (Machinist, *The Epic of Tukulti-Ninurta I: A Study in Middle Assyrian Literature,* 66–70, 110–112, 118–120, respectively; and cf. the discussion there, 151–155, 352). A similar example details the wickedness of the inhabitants of Babylon, which led Marduk to abandon them, destroy the city and give its inhabitants over to Sennacherib; see Borger, *Asarh.,* pp. 12–14. It is unlikely that theological arguments of this type were useful or convincing in practice.

> after our withdrawal, and reflect many times in your deliberations that your fatherland is at stake, your own and only fatherland, and that upon one decision only will depend her fate for weal or woe (*ibid.*, 111).[41]

In a similar incident, Ben-hadad demands total submission of Ahab and emphasizes his own great strength: "May the gods do thus to me and even more, if the dust of Samaria will provide even a handful for each of the men who follow me" (1 Kgs. 20:10).

The aggressor's determination is also reflected in the message sent by Pi(ᶜanhk)y to Mer-Atum (Meidum) "Look, two ways are before you; choose as you wish. Open, you live; close, you die. My majesty will not pass by a closed town!" (Pi('anhk)y Inscription, line 82).[42] This is also reflected in similar circumstances, in Jeremiah's comments (as King Nebuchadnezzar of Babylon is perceived as executing God's plan):

> And to this people you shall say: Thus said the Lord: I set before you the way of life and the way of death. Whoever remains in the city shall die by the sword, by famine, and by pestilence; but whoever

41 A negotiation conducted close to the beginning of a siege, with recognizable lines of similarity to the biblical passages concerning Rabshakeh's delegation and the Melian dialogue, appears in Esarhaddon's "Letter to the God" (Borger, *Asarh.* §68). After Esarhaddon's entreaties to King Ik-Teshub of Shubria, requesting him not to shelter political refugees who were rivals of the Assyrian king, had been rejected three times, the Assyrian army attacked the land of Shubria. As the Assyrian army approached Uppume, the capital of Shubria, Ik-Teshub made an unequivocal demonstration of total submission and begged for forgiveness. But all his entreaties were rejected by Esarhaddon, who emphasized that at that stage, there was no longer room for pardon. The city came under siege and was conquered, and other towns in the land of Shubria were also destroyed. On the political and ideological background of the negotiation between Esarhaddon and Ik-Teshub, see Oppenheim, in *Propaganda and Communication in World History,* I: 111–144, esp. 123 ff.; Leichty, in *Ah Assyria* ..., 52–57; see also Ephᶜal, *JCS* 57 (2005), 99 ff. This negotiation was unique, in that the side which proposed total submission was the party under attack, while the aggressor was determined in his decision not to reach any arrangement other than conquest of the city and its destruction, see Ephᶜal and Tadmor, in *Essays on Ancient Israel in Its Near Eastern Context*, 163–168.

42 Grimal (above, p. 45, n. 32), 88; Lichtheim, *ibid.*, 74

> leaves and goes over to the Chaldaeans who are besieging you shall live; he shall at least gain his life. For I have set my face against this city for evil... (Jer. 21: 8–10).

When demands for submission were accompanied by an ultimatum, the threat became even stronger. The following message was sent by Pi(ʿanhk)y to the besieged town of Per-Sekhemkheperre ("House of Osorkon I"):

> O you who live in death, you who live in death; you poor wretches, you who live in death! If the moment passes without your opening to me, you will be counted slain according to the king's judgment. Do not bar the gates of your life, so as to be brought to the block of the day! Do not desire death and reject life! (ll. 76–78).[43]

One may conjecture that the effectiveness of such declarations depended on previous successes of the threatening party.[44]

The psychological pressures on the besieged continued throughout the siege. A relief from Sargon's palace at Khorsabad depicts an Assyrian attack on the city of Pazashi located at the border of Mannaea; in it, an Assyrian officer is shown standing at the window of the turret of a battering ram positioned opposite the gate of the city, reading from a scroll held in his hand. According to Yadin, this officer, like Rabshakeh, Sennacherib's emissary to Jerusalem, is probably reading a demand for the surrender of the besieged city.[45] The location of the ramming machine opposite the city gate, on the road leading to it (and not on the ramparts, like another machine in the same relief, to the left of the city), indicates a stage of negotiations, while the presence of the officer in a military machine that shields most of his body, and the burning torches falling on the ram, indicate the rejection of his appeal.

43 Grimal, *ibid.*, 80; Lichtheim, *ibid.*, 74

44 See the descriptions of Pi(ʿanhk)y's successes in Hnes (=Heracleopolis; ll. 17ff.), in Permedjed (=Oxyrhynchus; l. 27), in the fortress of "the Cliff of Great Victories" (ll. 27–28), in Hut-benu in the nome of Ankyrononpolis (=El Hiba; l. 29), and in Hermopolis (ll. 31ff.).

45 See Yadin, *Art of Warfare,* 319–320, 425. In light of the status of Aramaic within the Assyrian Empire, it seems likely that the scroll was written in that language. See Tadmor, *EI* 20 (1989), 249–252.

Fig. 4: Assyrian attack on the city of Harhar in Media: relief from Dur-Sharrukin (Khorsabad). Below, people impaled in the course of combat over the city
From Botta & Flandin, *Monument de Niniveh, I/II, Pl. 55*

A cruel method, intended to break the spirit of the besieged and force them to surrender, was the torture and execution of their relatives before their very eyes. The annals of Tiglath-pileser III describe his war against Damascus as follows:

> His chief ministers (i.e., those of Rezin of Damascus, in the year 733 BC) I impaled alive and had his country behold them. For forty-five days I set up my camp around his city, and I cooped him up like a bird in a cage. I cut down his gardens… and orchards without number; I did not leave a single one.[46]

46 *ITP* 78:9'–11'.

It is also possible that the impaled figures on the Assyrian reliefs of siege battles were put to death *before* the city fell, to weaken its resistance, rather than as a demonstrative punitive measure after its conquest, "that they may see and fear."[47] Aeneas the Tactician (10.23–25) raises the possibility that inhabitants of a besieged city may cooperate with the enemy when their relatives are captured by the attackers, and may be subject to torture and death.

Various sources mention the cutting down of fruit trees in connection with siege on cities that have not been conquered. Thus, for example, the annals of Ashurnasirpal II of Assyria describe his campaign against Ilanu, ruler of Bit-Zamani, during which the fortified city of Damdamusa was captured in a siege battle. The annals state that the Assyrian king took four hundred combatants captive (and three thousand civilians). These captives and the leaders of the cities were taken to Amedu, the royal city of Ilanu, where their heads were placed in a pile before the gate of the city. These soldiers were impaled alive on stakes around the city; a battle took place at the city gates and its orchards were cut down, but it is not explicitly stated that the city was captured.[48] A similar scene is found in the inscriptions of Shalmaneser III, in the description of the defeat of King Hazael of Damascus: "To save his life he ran away (but) I pursued him. I imprisoned him in Damascus, his royal city, (and) cut down his gardens" (again, the text makes no explicit mention of the city's conquest).[49] The cutting down of fruit trees is also mentioned in the description of the battle of Tiglath-pileser III against Damascus, quoted above, and depicted in the reliefs of Shalmaneser III and Sennacherib.[50]

47 On the depiction, in the reliefs, of figures impaled during an attack on a city, rather than as a punitive measure after its conquest, see Yadin, *Art of Warfare,* 407, 424–425.

48 RIMA 2 220.

49 RIMA 3 48:14"–17".

50 The gates of Balawat, B. O. 2, in Billerbeck & Delitzsch, *Die Palasttore Shalmanassars II. von Balawat,* Tafel 1; Paterson, *Assyrian Sculptures: Palace of Sinacherib,* pl. 13. The cutting down of fruit trees in connection with war against a city is also depicted on a silver bowl from Amathus (Cyprus), dating from ca. 725–625 BC. See Barnett, *RDAC* (1977), 164–169.

The cutting down of trees causes long-term damage that lasts for many years (unlike the destruction of fields, which is limited to one year only). From these descriptions, it appears that the purpose of cutting down trees was not to leave the enemy's land in desolation as an end in itself,[51] but can be seen as another method to pressure and break the defenders' spirits (like the impalement of people in the presence of their relatives in the city and on the wall). Hence, it seems plausible to assume that this act was not carried out all at once, but done piecemeal, over an extended period, in order to increase the pressure on the besieged and to force them into negotiations and surrender. This analysis regarding the ongoing nature of this activity may clarify the meaning of the verse, "When you lay siege to a city *for many days* and make war against it to capture it, you should not destroy its trees" (Deut. 20:19).[52]

(c) Rebellion or Riots in the City

A city's ability to oppose a siege depended on the solidarity of the people within it (especially as not all of these were permanent inhabitants; see below, pp. 162–167). Therefore, authorities needed to take precautions against a "fifth column" within the city. Aeneas the Tactician gives instructions for the designation of loyal groups to handle the opposition (1.6), for steps intended to prevent the gathering of rebellious groups in open areas within the city and methods of controlling them (2.1, 7–8), and for the prevention of any rebels from communicating with the besieging enemy using light signals at night (10.25–26).

51 On the cutting down of orchards of fruit trees simply to increase destruction, see, e.g., 2 Kgs. 3:18–19, 24–25; Sargon's "Letter to the God Ashur" concerning his eighth campaign, TCL III 42:265–267; 44:276; 46:296, 303. On the literary model for the description of destruction caused by Sargon during this campaign, which is the opposite of the image of the king who promotes growth and flourishing, see Zaccagnini, in *Assyrian Royal Inscriptions: New Horizons in Literary, Ideological, and Historical Analysis*, 259–295; see also Cole, in *Assyria 1995*, 29–40 (with additional examples).

52 The conclusion that the text refers to damage to fruit trees regardless of military benefit is strengthened in light of the fact that of all fruit trees growing in Palestine, only the date palm can be used for construction.

The dissemination of defeatism and calls to surrender were considered a crime punishable by imprisonment or even death.

> Shephatiah son of Mattan, Gedaliah son of Pashhur, Jucal son of Shelemiah, and Pashhur son of Malchiah heard what Jeremiah was saying to all the people: "Thus said the Lord: Whoever remains in this city shall die by the sword, by famine, and by pestilence; but whoever surrenders to the Chaldeans shall live; he shall at least gain his life and shall live.[53] Thus said the Lord: This city will surely be delivered into the hands of the king of Babylon's army, and he shall capture it." Then the officials said to the king, "Let that man be put to death, for he disheartens [lit., weakens the hands of] the soldiers, and all the people who are left in this city, by speaking such things to them. That man is not seeking the welfare of this people, but their harm!" King Zedekiah replied, "He is in your hands; the king cannot oppose you in anything!" (Jer. 38: 1-5).[54]

Opposition groups of various kinds could operate in besieged cities, due to internal reasons bearing no direct relationship to the besieging enemy; however, there are indications that the besiegers often took steps to drive a wedge between the city rulers and any elements in the population for whom resistance to the enemy was not vital. Stirring up these elements could lead to the surrender of the city without expensive military effort.

A similar approach, sowing mistrust in the leaders among the population, with the intention to stir up social unease, is implied in Rabshakeh's appeal "to the men sitting on the wall (i.e., the combatants)" in Jerusalem (2 Kgs. 18:28-35). His comments, made "in a loud voice in Judean" – that is, in the language of the besieged population rather than in Aramaic, the language of choice for King Hezekiah's representatives – Rabshakeh points out that

53 Note that this does not imply that one who goes out to the enemy will not be exiled; compare similar phrasing by Rabshakeh, "Make your peace with me and come out to me... until I come and take you away to a land…" (2 Kgs. 18: 31-32).

54 There is no need to adopt the approach of those exegetes who find in Zedekiah's comments evidence of his weak position relative to the officials. The literal meaning, that the king could not reject a practice required by the dire situation of the besieged city, is to be preferred.

it is the population who are forced "to eat the dung and drink the urine." In addition to the literal meaning of this statement, the text can be interpreted to mean that the combatants and the people would have to pay for their monarch's political adventure. His words are directed explicitly against Hezekiah's words of encouragement: "Do not let Hezekiah deceive you" (v. 29); "Do not let Hezekiah make you rely on the Lord…" (v. 30); "Do not listen to Hezekiah" (v. 31); "Do not listen to Hezekiah, who misleads you" (v. 32). Fearing similar behavior, the leaders of Melos refrained from bringing the representatives of Athens (who sought their surrender) before a public gathering, but requested that they declare the purpose of their visit before a limited circle only, including the city leaders and probably the chief governing body. The Athenians' reaction was:

> Since our proposals are not to be made before the assembly, your purpose being, as it seems, that the people may not hear from us once for all, in an uninterrupted speech, arguments that are seductive and unattested (i.e., not questioned or put to the proof), and so be deceived, for we see that it is with this thought that you bring us before the few (Thucydides, 5.85).[55]

The emissaries of Tiglath-pileser III made a similar address to the citizens of Babylon (see above, n. 33) during the rebellion of their king, Ukin-zer the Chaldean, declaring that the Assyrian king would honor their right of *kidinnūtu* (NL 1).[56] Likewise, Joab explains to the wise woman in Abel-beth-maacah that he does not intend to destroy the city but only to take Sheba son of Bichri, who was protected by the city's leadership even at a very advanced stage of the siege.[57] His explanation produced the intended result:

55 Compare, similarly, Aeneas the Tactician's instructions to restrict the opportunities for direct communication between representatives of states, armies and other foreign groups with the population of a city besieged or in danger of siege (10.11).

56 A similar tendency is reflected in the proclamation sent by Ashurbanipal to the inhabitants of Babylon during his struggle with his brother, Shamash-shum-ukin (*ABL* 301). However, it should be noted that the Assyrian king's appeal was made on 23 Iyyar 652 BC, while the city did not come under siege before 11 Tammuz 650 BC (Grayson, *Chron.* 15:19).

57 This advanced stage is reflected in the activities undertaken by the soldiers of Joab

The woman came to all the people [did this include the city leaders or had a rebellion broken out in the city? – IE] with her wisdom; and they cut off the head of Sheba son of Bichri and threw it down to Joab. So he sounded the horn, and all the men dispersed to their homes, and Joab returned to the king in Jerusalem (2 Sam. 20:22).

2. Famine

The primary and most common result of blockade is famine, which is intended to break the spirit of the besieged population and to put an end to their ability to resist.

Upon the approach of the enemy, the reaction of the population, particularly that which lived outside of these cities, was: "assemble, and let us go into the fortified cities" (Jer. 4:5; 8:14).[58] The nature of these cities as sites for the storage of food and of war materiel is shown in 2 Chron. 11:11–12. Following the list of "fortified towns" (ערים למצור / ערי המצורות) built by Rehoboam in Judah, we read: "He strengthened the fortresses and placed in them commanders, and stores of food, oil, and wine; and in every town he put shields and spears, and strengthened them exceedingly."

The population congregating in the fortified city included the inhabitants of unwalled settlements in the area, semi-nomadic populations and armies who withdrew to the city after a field battle had been conducted in the vicinity.[59] The concentration of a large population in the besieged city

---

and David: "[Joab's men] came and besieged him (Sheba son of Bichri) in Abel-beth-maacah, raised a siege ramp against the city… All the troops with Joab were engaged in battering the wall" (2 Sam. 20: 15).

58 Compare, "I will bring a sword against you to wreak vengeance for the covenant; you shall withdraw into your cities" (Lev. 26:25); *mātu ana dannati ipaḫḫur,* "The land will [have to] assemble in a fortress," Leichty, *Izbu* xi 74'; *nakru māta ana dannati upaḫḫar,* "An enemy will cause the land to assemble in a fortress," Nougayrol, *RA* 65 (1971), 73:25'. On the gathering of the populations of scattered villages within a fortified city due to the approach of the enemy, see ARM XXVI 515:29–33; XXVII 113:11–15.

59 See below, pp. 162–167. Although the sources at our disposal do not elaborate on the size of the besieged populations and the besieging armies, several aspects merit attention. The area of Lachish at the time of its conquest by Sennacherib was about 7.5 hectares. Assuming the density index in the city at the time of the siege was 400

suited the goals of the attackers. In his directives to military commanders, Onasander (first half of the first century CE) suggests concentrating all non-combatant population outside of the city into the city itself, in order to increase the depletion of provisions available to the besieged.[60]

Shortage of fodder, barley and even water prevents the feeding of animals in a besieged city for a long period, forcing the besieged to slaughter their animals early in the siege. Thus, we find that the beginning of a siege was a period of gross consumption of flesh, which created a particular atmosphere.[61] This mood is expressed in Isa. 22:12–13, describing the agitation in Jerusalem in anticipation of the approaching disaster (vv. 1–4), the approach of the enemy (i.e., the Assyrian army), the enemy's preparation for war against the city (vv. 5–8a) and the extensive work of fortification performed within the city in preparation for siege (vv. 8b–11), as a result of which the prophet anticipates "weeping and lamenting, tonsuring and girding with sackcloth" (v. 12).[62] But the actual behavior of the besieged was completely different: "Instead, there was rejoicing and merriment, killing of cattle and slaughtering of sheep, eating meat and drinking wine: 'Eat and drink, for tomorrow we die'" (v. 13). This is not necessarily festivity or hilarity, but rather a natural behavior of people under siege; like many of the ancient Near East, they did not ordinarily eat much meat except at festive meals; now, due to the circumstances (and

---

people per hectare (approximately twice the regular density index in built-up areas in Palestine and in the Near East during the period discussed), it follows that the besieged population was approximately 3,000. According to this density index at the time of Nebuchadnezzar's siege of Jerusalem, there were approximately 26,000 people (=400 x 65) in the city. The length of Jerusalem's walls at that time was approximately 3,250 meters. The sizes of the besieging armies should also be estimated accordingly.

60 Onasander 42.23; compare 1 Macc. 6:53.

61 An additional factor that frequently increased excessive eating was the absence of suitable means of meat preservation for more than a day or two. Hence it was usual to gorge on the meat, at times even in exaggerated quantities, before it became spoiled. A shocking example of this reality is found in the story of the two women in besieged Samaria who agreed between them to consume the flesh of their two sons within two days (2 Kgs. 6:28–29), a quantity far greater than the consumption ability of two women.

62 For such behavior in a besieged city, cf. Ezek. 7:18.

even despite themselves!), they gorged on large quantities of meat. In such an atmosphere, the expected disaster and apprehension are pushed aside and suppressed.

As the siege continued, and the supply of food within the city diminished,[63] the people needed to find substitute forms of food from whatever was available. To symbolize the food shortage expected in Jerusalem during the siege, Ezekiel was commanded to bake a single cake made of wheat, barley, beans, lentils, millet and emmer (Ezek. 4:9). This peculiar combination certainly tasted very different from ordinary barley cake (compare v. 12), and has been decreed unfit even for a dog (*b. Eruvin* 81a); however, there is no doubt that in a time of adversity and want, there would have been many who would have been happy to eat it.[64] The same prophecy emphasizes the shortage of food by the limited quantity allotted to the prophet: he is to eat "by weight, twenty shekels [about 320 grams, according to the standard of the heavy Babylonian shekel] a day; once a day" (v. 10).[65]

63 In this connection, one should note the appreciable waste in grain and other foods stored for extended periods. During a two-year siege (such as Ashurbanipal's siege on Babylon or Nebuchadnezzar's siege on Jerusalem), no less than a quarter of the grain stored in silos and in houses was ruined due to insects and other vermin. See Adamson, *WO* 16 (1985), 5-15.

64 For a typological parallel – the mixing, due to famine, of flour with foods ordinarily eaten by wild donkeys, see *ABL* 1000:7-11: *ina qēmi ša murrutu u zēri ša sungirti ša sirrīmū ikkalū kabêšunu iḫaššalūma inappûma ana libbi aḫāmeš usammaḫūma iṭennūma ippû ikkalû ina libbi balṭū*, "They live off the flour of *murrutu* (a kind of plant) and seeds of *sungirtu*, eaten by wild animals: they pulverize the husks (of the seeds), winnow them and mix them, grind them, bake them and eat them" (thus *CAD*, esp. K 29 *kabû* C). Von Soden (esp. *AHw* 419a; *kabūtu(m)* 3; 1387 *ṭênu*) translates the word *kabû* as "dung" (this is its usual meaning, compare *CAD kabû* A). The meaning of the sentence according to this translation would then be: "They eat the flour of *murrutu*; in addition, they crumble the dung of wild beasts (and gather from them) the seeds of *sungirtu*, etc." (on the gathering of barley from animal dung during famine, cf., e.g., *b. Ketubot* 66b).

65 On lack of food in the besieged city, compare: "(You shall be delivered into enemy hands) When I break your staff of bread, ten women shall bake your bread in a single oven; they shall dole out your bread by weight, and though you eat, you shall not be satisfied" (Lev. 26:26). Also see: "(King Zedekiah) gave instructions to lodge Jeremiah in the prison compound and to supply him daily with a loaf of bread from the Baker's Street – until all the bread in the city was gone" (Jer. 37:21).

The atmosphere of misery and deprivation associated with the eating of bread "by weight" and the drinking of water "by measure," is characterized as "anxiety" and "horror" (Ezek. 4:16).[66] Under such harsh conditions, people tried to ease the gnawing pangs of hunger by various forms of distraction. One example of such a distraction is cited by Herodotus (1.94) in the name of the Lydians, who are said to have suffered from a protracted famine (18 years, due to drought) and devised various games involving dice, bones and balls to distract themselves; one day they played, and as a result did not think about food, and on the second day they refrained from playing and ate.

In the course of time, even acceptable food substitutes disappeared, and people no longer rejected anything whatsoever. Among the foul substances eaten in times of siege, literary sources mention "a donkey's head" and "dove's dung" (*ketib:* חרייונים; *qeri:* דביונים), which were sold at an exorbitant rate (in Samaria, see 2 Kgs. 6:25),[67] and leather straps (*kurussu*) which

---

The protracted siege also involved a shortage of kindling material for baking and cooking. Even in ordinary times, most cities of the Near East were not blessed with an abundance of firewood and the storing of wood for times of siege was considered particularly difficult due to its great volume (compare *b. Gittin* 56a concerning the three rich men of Jerusalem during the time of Vespasian's siege who committed themselves to provide for the entire city: one with wheat and barley, one with wine, salt and oil, and one with wood. "The Rabbis considered the offer of wood most generous... for R. Hisda would say: 'A storehouse of wheat requires sixty stores of wood [for fuel].'" Baking ovens were generally stoked with cow dung (צפיעי הבקר; Ezek. 4:15). However, this form of fuel disappeared with the decrease in the numbers of animals in the besieged city, and Ezekiel was commanded to bake his loaf of bread with "the dung of human beings" (vv. 12, 15).

66 Compare the eating of bread "with noise" and "with worry" and the drinking of water "with anger and with anxiety and with desperation" (Ezek. 12:18–19).

67 The meaning of the expressions ראש חמור ("a donkey's head") and חרייונים is not entirely clear. Some scholars argue that these terms designate the names of certain plants – see Oppenheim, *JQR* 37 (1947), 175–176; Held, *Studies in Honor of B. Landsberger,* 395–398; and cf. CAD Ḫ 43 *ḫalla;* S 380 *summatu,* 3. According to Held, the phrase חרי יונים is equivalent to חרובים and is to be identified with *Ceratonia siliqua L.* (carob; Akk. *ḫarūbu;* Hebrew חרוב), or with *Prosopis stephaniana* (false carob, Hebrew ינבוט, Akkadian *ašāgu,* Arabic *ḫarnūb*, *šauk*). Both species grow in Palestine. The carob, which is native to the same approximate region as the olive, does not grow in the alluvial areas of Mesopotamia. On the archaeo-botanic remnants indicating that the

were chewed by the famished (in Babylon during Ashurbanipal's siege).[68]

Literary sources reflect the damage caused to the body by inadequate nutrition and starvation: people's skin became shriveled and darkened (Lam. 4:7–8), the milk of nursing mothers dried up (*ibid.* 4:3–4),[69] and people were greatly weakened by the "sickness of famine" (Jer. 14:18). Descriptions of people lying about in the streets of Jerusalem in a faint due to hunger appear in Isaiah (51:20) and in Lamentations (2:11–12; 4:5). If no help came, the process concluded in a long, painful death: "Better off were the victims of the sword than the victims of hunger, who pined away, stricken, by want of the fruits of the field" (Lam. 4:9).[70]

One of the motifs found in the *topos* of famine in ancient Near Eastern sources is that of cannibalism, which reflects the serious psychological effects of hunger to the point of loss of humanity. This motif is widespread in the omen literature, in political treaties and in various kinds of literary works.[71] The history of the human race includes extreme cases in which

---

carob grew in Palestine during the 1st millennium BC and even before that (albeit rarely), see Liphschitz, *Israel Journal of Botany* 36 (1987), 191–197. The carob was considered animal food; its consumption by humans was considered a sign of poverty and hardship. For other interpretations of these expressions, which do not resolve the difficulties, see Tsukimoto, *AJBI* 12 (1986), 77–86.

68 Streck, *Asb.*, 36 iv 44–45; compare Borger, *Beiträge zum Inschriftenwerk Assurbanipals*, 150: "(viii 112) (The inhabitants of Babylon) chewed hide and straps (113) of shoes and sandals (*ik-su-su gi-il-du ku-ru-us-su* ^{kuš}DA.E.SIR.MEŠ *u* ^{kuš}E.SÍR.MEŠ). (114) In their hunger they slaughtered (115) sons, daughters, brothers, sisters (116) …all of them."

69 For discussion of the lack of milk for infants as a common motif in ancient Near Eastern sources (the Sefîre treaty, Ashurbanipal inscriptions and the Bible) in connection with suffering and hunger, see Hillers, *Treaty-Curses and the Old Testament Prophets*, 61–62.

70 For discussion of the bodies of victims of starvation and plague that were cast into the streets without burial, see below, pp. 129–135.

71 The following references deal explicitly with cannibalism during siege: Lev. 26:29; Deut. 28:53–57; 2 Kgs. 6:28–29; Jer. 19:9; Ezek. 5:10; Lam. 2:20; 4:10; Borger, *ibid.* (note 68), 150 viii 117–120; Streck, *Asb.* 36 iv 44–45 (the Assyrian passages cited here relate to Babylon).

Political treaties record parents eating their offspring without any explicit mention of siege, see SAA II 6:449–450, 547–550, 568–569 (Esarhaddon's succession treaty). On references to the consumption of human flesh, not necessarily that of children,

people resorted to cannibalism; however, cannibalism of one's offspring is evident as a distinctive literary motif in ancient Near Eastern sources relating to siege.[72] This horror finds its severest expression in biblical literature: "Look, O Lord, and see! With whom hast thou dealt thus? Should women eat their offspring, the children of their tender care? Should priest and prophet be slain in the sanctuary of the Lord?" (Lam. 2:20); "The hands of compassionate women have boiled their own children; they became their food in the destruction of the daughter of my people" (Lam. 4:10); the father consumes the flesh of his sons and daughters without sharing it with his brothers, with the wife of his bosom, or with his remaining children; the mother eats the flesh of her children in secret, out of fear that her husband, her sons or her daughters will snatch the flesh out of her hands (Deut. 28:53–57); and in the description of the two women who make an agreement to share the flesh of their children, one each day (2 Kgs. 6:28–29).

---

in the omen literature, without explicit mention of siege, see, e.g., *ACh* Adad XVII 36; Suppl. LIX 15; Sin XXV 16; Šamaš X 5. It would seem that at least some of the references to cannibalism may be connected with the severe famine caused by siege. On *ACh* Sin XXXIII 76 and 2 Kgs. 25:3–6 // Jer. 52:6–9, see above, pp. 18–19.
The formulation of the motif of cannibalism in the treaties and in at least some of the literary sources displays a clear relationship: the view that the necessity of eating one's relatives is a punishment explicitly mentioned in the covenant, imposed on those who have violated a solemn oath, occurs in Deut. 29:23 ff. (following the imprecations of Deut. 28:15–68); Streck, *Asb.* 76–78 ix 59–64, 68–74 (see also Greenberg's remarks in the following note).

72 Two of the literary sources listed in the previous footnote are exceptions to this rule: (a) the author of the Rassam Cylinder used the motif of cannibalism in describing the miserable situation of the city of Babylon during Ashurbanipal's siege. While he wrote of the inhabitants of Babylon – in accordance with the standard formula – that they ate the flesh of *their children* (Streck, *Asb.*, 36 iv 43–45), he took care to mention that the Arabs, who found themselves in the city by chance during the siege, ate *one another's* flesh (*ibid.*, 68 viii 35–37) – quite correctly, because the latter came to Babylon as warriors who had come from the desert (without their families) to help Shamash-shum-ukin, and therefore it was impossible to write that they ate the flesh of their children (see Ephʿal, in *Sefer Moshe*, 28–30); (b) in Ezek. 5:10 it states "Therefore, fathers shall eat their sons in the midst of you and sons shall eat their fathers." On the relationship of this verse (and cf. *ibid.*, v. 9) to Deut. 24:16 and its fashioning, see Greenberg, *Ezekiel 1–20,* 113–114.

Crops in Palestine in Biblical Times

The Crop	The Period of Ripening
Barley	April-May
Wheat	May-June
Millet	June
Rye	May-June
Broad Beans	April-May
Lentils	June-July
Sorghum	August
Figs	June-August
Almonds	August-November
Grapes	September-November
Dates	September-November
Olives	August-December

From the attacker's viewpoint, the most suitable time to begin a siege was some time before the crops ripened, when the old year's supply of food was depleted and no fresh produce had been gathered yet to take its place. A particularly fortuitous time for beginning a siege against the Jews in Palestine occurred once every seven years, at the close of the sabbatical year (מוצאי שביעית ; there is evidence of the observance of the sabbatical year during the Second Temple period and thereafter), that is, during the months of Tishri through Nisan (roughly October through April) of the *eighth* year, during which the harvest of the sixth year had been completely consumed and the crops of the eighth year had not yet ripened.[73]

We have little information concerning the dates of the beginning of siege: (1) the 11th of Tammuz in the eighteenth year of Shamash-shum-ukin (July

73 This was the situation during the campaign of Lysias, captain of the army of King Antiochus V of Syria against Judah in the year 162 BC, compare 1 Macc. 6:49, 53–54.

651 BC) marked the beginning of Ashurbanipal's siege of Babylon;[74] (2) the 10th of Tebeth in the ninth year of Zedekiah's reign (that is, 15 January 588 or 5 January 587 BC) signaled the beginning of Nebuchadnezzar's siege of Jerusalem.[75]

Ashurbanipal's siege of Babylon began at the ideal time for a siege in terms of the agricultural calendar – the exact period at which the produce had ripened, but before the people could manage to gather it (or at least most of it) and bring it inside the city (a certain period of time was also required for threshing and winnowing).[76] On the other hand, it is clear that the beginning of the siege of Jerusalem was not determined on the basis of the seasons of gathering and replenishing food supplies by the besieged (from a logistic viewpoint, this was also inconvenient for the attackers), but rather according to political circumstances and other military considerations.

3. Thirst

While famine is frequently mentioned in the Akkadian queries, as well as in other sources from the ancient Near East, references to defeat of a city through thirst are unusual.[77] This is hardly surprising: famine does not leave its mark until after an extended period of time, whereas the influence of dehydration is visible within a few days.[78] It therefore seems implausible

74 Grayson, *Chron.* 15:19

75 This date is calculated on the basis of 2 Kgs. 25:1; Jer. 39:1; 52:4; cf. Ezek. 24:1 ff. For the debate on the duration of Nebuchadnezzar's siege against Jerusalem – i.e., whether it lasted 18 or 30 months – see Malamat, *IEJ* 18 (1968), 150 ff.

76 In Babylonia of the 6th and 5th centuries BC, the barley and wheat harvest took place during the months of Iyyar and Sivan, see Neumann and Sigrist, *Climatic Change* 1 (1978), 239–252. On the calendar of agricultural tasks in Babylonia for the months from Iyyar through Elul, see also el-Sāmarrāie, *Agriculture in Iraq during the 3rd Century, A. H., 66–68.*

77 SAA IV 102:6; IM 67692:59.

78 Ezekiel was commanded to drink water in measure "a sixth of a *hin* (i.e., 0.6 liter) each day" (Ezek. 4:11). Such a small ration of water is sufficient only as a symbolic act, and cannot be maintained for more than a few days. Water shortage leaves the besieged no option, forcing them to surrender quickly (cf. Jdt. 7:7, 12–13, 20–28). Cf. SAA XV 189 rev. 10'–13e: "There is no water in Dur-sharrukku. If you come and launch an attack on it, you will take it in a matter of a day." See also Joab's letter to David: "I have attacked

that a city lacking sufficient water reserves would choose to endanger itself through siege (see the theoretical notes above, pp, 1–3). If such resources (a river or a canal flowing through or close to the walls,[79] a spring or protected water system within the fortified area,[80] or at least hewn cisterns for water storage[81]) were unavailable, it would have been preferable to surrender

---

Rabbah (Rabbath-ammon) and I have already captured the city's water. Now muster the rest of the troops and besiege the city and capture it; otherwise I will capture the city myself, and my name will be connected with it" (2 Sam. 12:27–28).

79 Most of the cities of Mesopotamia and Egypt enjoyed the waters of rivers or canals that flowed through them or close to them and were very useful in times of calm. However, in times of war, the canals could become a disadvantage, as they could easily be blocked downriver from the city, thereby flooding it (see above, p. 21, n. 43), or diverted, thereby removing the city's water supply, and even used to penetrate the city through a dry riverbed under the wall. In this respect, water systems or cisterns within the city, out of enemy reach, are an advantage.

80 See, for example, the description of water systems discovered in various sites in Palestine (including Hazor, Megiddo, Gezer, Gibeon, Jerusalem and Lachish), dated to the 10th–8th centuries BC: Shiloh, in *The Architecture of Ancient Israel from the Prehistoric to the Persian Periods,* 275–293. A relief found in Ashurnasirpal II's palace in Calah (Nimrud), BM 118906, depicts an Assyrian soldier cutting a rope tied to a bucket that is lowered down from the wall of a besieged city with a wheel (see Yadin, *Art of Warfare*, 460–461). Albenda argues that this relief shows a technique of drawing water within the city and not outside of it (so, this is a water system dug within the city) that was customary in Syria and in Palestine as early as the end of the second and beginning of the first millennium BC. Accordingly, she connects the relief in question to Ashurnasirpal II's campaign in Syria and Phoenicia; see Albenda, *BASOR* 206 (1972), 42–48. If Albenda's hypothesis is correct, it is impossible to infer from the relief how the Assyrian soldiers could have harmed the water system before entering the city and conquering it.

81 The Mesha Inscription, ll. 21–26, enumerates Mesha's activities in Dibon: "(21) I built for the citadel (קרחה) the 'wall of the forests' and 'the wall of (22) the rampart' and I built its gates and I built its towers and (23) I built a royal palace and I made the channels or the reservoir for water in the mid(24)st of the city. But there was no cistern in the midst of the city, in the citadel, so I said to all the people 'Make [for] (25) yourselves each man a cistern in his house.' And I hewed the shafts for the citadel with prisoner(26)s of Israel." Mesha is not praising his activities for the benefit of the inhabitants of Dibon; on the contrary, the order to "make for yourselves each man a cistern in his house" involved considerable burden on the population. But, the matter is understood differently in the context of preparations for the fortification of Dibon and making it a "fortified city."

from the outset.[82] Consequently, even scattered references to thirst in the queries should not be interpreted as referring to an ordinary siege of a city, but to special circumstances in which the central water source of a city was seized in the course of battle, as a result of which the city was subject to thirst and quickly defeated.[83]

4. *Epidemics*

Another danger that accompanies sieges is the epidemic, which breaks out as the result of inferior sanitation, crowded conditions, water contamination and an inadequate and unbalanced diet.[84] Thucydides (2.47–54) describes in detail the plague which struck Athens and Piraeus during the invasion of Attica by the Peloponnesian alliance in 430 BC, in whose wake the refugees from that country fled to these two cities.

Plague is one of the three characteristic afflictions of war, and particularly of siege, in the literary sources: plague (Hebrew דבר/מות; Akkadian *mūtānu;* Aramaic מותנ[א]), famine and the sword.[85] It should be noted that famine and

82 An exception to this is Tyre, which was located on an island without any sources of running water and was dependent on supplies brought in on boats. Every siege cut Tyre off from the water sources available on the adjacent shore. The Tyrians overcame this difficulty by bringing water from a greater distance, thanks to their naval ability, which was superior to that of their enemies; over time, they also hewed cisterns which made it easier to sustain interruptions in the regular supply of water. See above, pp. 40–41, nn. 21–23.

83 Compare the omens: CT 30 45 83-1-18,415 rev. 11: *nakru ālka ilammīma ina šiḫiṭ šamši, [...] būra liḫrâ mê būrēka ana būrē<šu> [..ālka] ina šimētān ina ṣūmi iṣabbat,* "An enemy will besiege your city, and he will dig a well at dawn... he will drain the water of your wells into his wells and capture your city at sunset by thirst"; *ibid.*, rev. 8: *nakru ālka ina šimētān ina ṣūmi iṣabbat*, "An enemy will capture your city at sunset by thirst."

84 Compare the remarks of Biridiya, the ruler of Megiddo, to Pharaoh concerning his plague-stricken city, following the siege imposed on it by the ruler of Shechem (EA 244:30–32).

85 See, for example, Jer. 21:6–9; 32:36; 38:2; 44:13; Ezek. 5:2, 12; 6:11–12; 7:15. On מות (lit. death) in the sense of plague, see: "And if they ask you, 'To where shall we go forth?' answer them, 'Thus said the Lord: Those destined for the plague, to the plague [אשר למות למות], those destined for the sword, to the sword; those destined for famine, to famine; those destined for captivity, to captivity'" (Jer. 15:2; compare also *ibid.* 43:11); "Give their children over to famine, mow them down by the sword. Let their wives be

the sword are extremely common in ancient Near Eastern sources relating to siege, while references to plague as a method of conquering cities are sparse in the historiographic works (the Bible and the Assyrian royal inscriptions),[86] and completely absent from the queries to the gods, which enumerate dozens of methods to achieve this goal.[87] This fact can be explained considering the conception at that time that famine and the sword are within human control, while disease and plague are exclusively reserved to the whims of the god(s).[88] One of the signs of this state of affairs is the fact that a plague could also afflict the attackers, regardless of their power or their activities.[89]

---

bereaved of children and husbands, let their men be struck down by the plague (הֲרֻגֵי מות) and their young men be slain in battle by the sword" (Jer. 18:21). See also the commentaries to Lam. 1:20: "Outside the sword deals, inside the plague." On *lipit Irra* as a euphemism for pestilence, see CAD L 201-202 *liptu* A,2,a.

86 See Ashurbanipal's inscriptions on his war against the cities of Babylonia, below, p. 131, nn. 49, 51.

87 The absence of references to plague in the queries seems to emphasize the practical nature of these documents – all of the numerous methods of conquering a city mentioned in these documents were in human control, and the documents were written expecting an answer regarding the choice of the best method.

88 On plagues as the weapon of the god(s), see: on Adad: RIMA 2 31:86 (inscription of Tiglath-pileser I); Nergal: the Tell Fekherye Inscription, Aramaic version, line 23; SAA II 6:455-456, and compare *ibid.* 4:26' (political treaties from the reign of Esarhaddon); the greater gods of Assyria: RIMA 2 242:27; SAA II 6:480; YHWH God of Israel: in addition to the biblical passages mentioned above in n. 85 and below in nn. 89, 90, see Exod. 5:3; Num. 14:12; Deut. 28:21; passim. On Hittite magical ceremonies intended to assuage the anger of the gods when a plague struck an army, see, e.g., Kümmel, *TUAT* 2/1 285-288; Sauček, *MIO* 9 (1963), 164-174.

89 See, for example, Diodorus on the plagues which struck the Carthaginian army when they made war against the cities of Sicily, Syracuse, in 413 BC (13.12.1, 4, 6) and Acragas in 406 BC (13.86.2-3); among the dead of that plague was the commander of the expeditionary force sent by Carthage against Sicily. According to Diodorus 13.114.2, more than half of the army of Carthage was lost in a plague during that campaign. On the dangers which awaited armies, see also the omen: "The disease of the *šību* shall take hold of (attack) the army" (YOS X 18:55).
The destruction of Sennacherib's army that laid siege to Jerusalem is described in the biblical legendary tradition as an angel of God striking the Assyrian camp (2 Kgs. 19:35 // Isa. 37:36). The same formulation – an angel striking the people – also appears in 2 Sam. 24, in the story of the pestilence (מגפה, vv. 21, 25) // plague (דבר, vv. 13, 15) that struck Israel following David's census.

The association of plague with siege or blockade appears in non-historiographic works.[90]

II. THE BREAKTHROUGH

Once the attackers decided to break into the city without waiting for the results of blockade, penetration could be achieved by three methods: through the wall, beneath the wall or over it. Each entailed a particular technique: penetration through the wall required battering rams, penetration beneath the wall required tunneling and breaking into the city over the wall required assault ladders. The basic characteristics of the three routes are detailed in the table below:

	Manpower Required	**Duration of Activity**	**Element of Surprise**
1. Assault Ladders	Small	Very brief	Yes
2. Tunneling in (tunnels)	Small	Relatively protracted	Yes
3. Breakthrough (ramparts and battering rams)	Large numbers	Extensive	No

The methods of breakthrough are listed here in the order of their preference from the attackers' standpoint. It seems plausible that the degree of urgency of the required counter-measures taken by the defenders followed the same sequence, so that steps were taken in advance to prevent or hinder

90 Lam. 1:20; see also the oracles of Jeremiah and Ezekiel, who were among the besieged in Jerusalem; Lev. 26:25: "I will bring a sword against you to wreak vengeance for the covenant; you shall withdraw into your cities, I will send pestilence among you, and you shall be delivered into enemy hands. When I break your staff of bread, ten women shall bake your bread in a single oven; they shall dole out your bread by weight, and though you eat, you shall not be satisfied."

penetration by ladders or tunnels. Consequently, at times, the attackers had no option but to attempt breakthrough using ramparts and battering rams – the most difficult of all methods of breakthrough.

The extant evidence suggests that in Mesopotamia and in the area subject to its technological influence, breakthrough was a common method of conquest from the first centuries of the second millennium BC onwards. By contrast, this method was not used in classical Greece until the end of the 5th century BC. Employment of this method of warfare seems to have been largely influenced by the political and social organization of the attacker: the time factor was of great importance in empires or in centralized royal systems, leading to the development of the methods of breakthrough. Apparently, the political organization of Greece, the polis system, was not conducive to this method.

1. Assault Ladders

The quickest and easiest way of breaking into a city was to go over the wall using assault ladders, a method which required no preparatory earthworks.[91] The use of assault ladders is documented in Egyptian and Assyrian reliefs, in Akkadian inscriptions, and in classical sources. Biblical references to this subject are discussed below.

This method of breakthrough required ladders suited to the height of the wall: if the ladders were too short, as happened to King Philip V of Macedonia when he attacked the city of Melitaea (217 BC), the entire campaign failed (Polybius 9.18.5, 8). Polybius suggests that the ratio between the length of the ladder and the height of the wall, the secant, should be 12:10. Thus, the angle between the ladder and the wall, whose secant is 1.2, is 35°. A secant less than 1.2, creating a more acute angle, would enable the defenders to easily push the ladder away from the wall (compare Aeneas the Tactician, Chap. 36).[92] A greater angle might cause

91 See above, p. 68.

92 According to Aeneas, if the ladder could not be repelled by hand, whether because of the cover of fire aimed at the wall or because the base of the ladder was too far away from the wall and the defenders did not have enough power to tilt the center of gravity of the ladder, including its human burden, to the other side of its point of support, a long forked pole could be used to push the ladder away or to injure those on it.

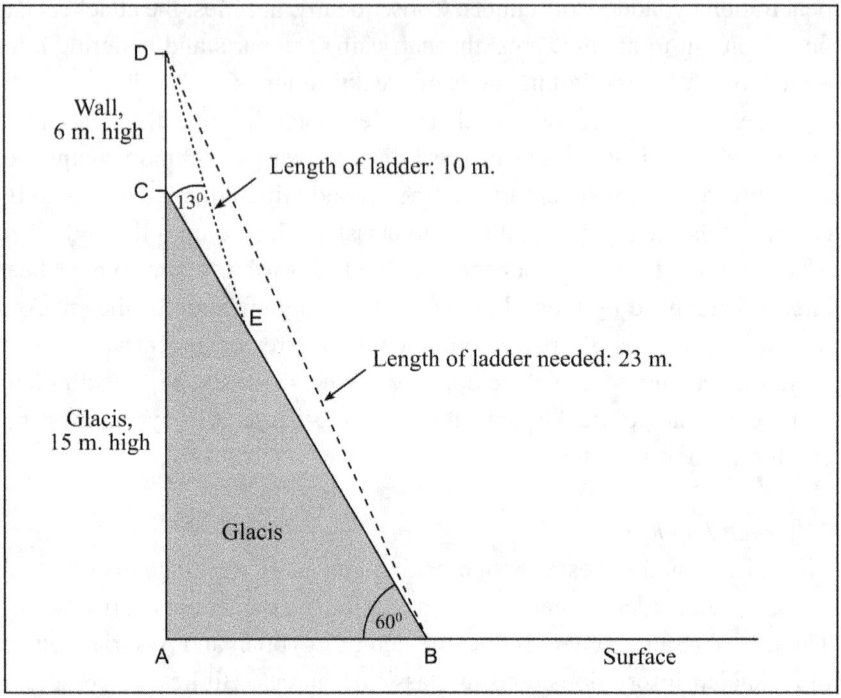

Fig. 5: The glacis as a means of defense against assault ladders

the ladder to break under the weight of the climbers. Polybius also explains how to make the geometric calculations to determine the length of the ladder required when it is impossible to get close enough to the wall to measure it; see 9.19.6–9).[93]

Since breaking into the city by assault ladders was, supposedly, relatively simple, it stands to reason that the defenders would take preventive measures against it. One of the basic methods of defense against invasion by ladders was the glacis, the sloped side of the wall which, in addition to other functions,[94] had a double effect on this method of attack: (1) it

93 On the calculation of the height of a wall by counting the layers of bricks in it from a reasonable distance, see Thucydides 3.20.

94 The glacis appears as a component of fortification in the ancient Near East from the 20th

increased the height of the wall, which was built above the glacis, thus considerably increasing the length of the ladder required, and (2) the glacis slope reduced the angle at the base of the ladder, thereby reducing its stability. Aeneas (36. 2) suggests another method for preventing ascent by ladders, namely, destabilizing the ladder at its base by a device prepared in advance, operated by a roller.

If the attackers succeeded in placing the ladders and leaning them against the wall, the defenders had only minutes to prevent the enemy from reaching the top of the wall. The placement of the ladders doubtless required a strong holding force to engage the defenders by fire and possibly also diversionary action and simultaneous attack at several points along the perimeter wall, to divert the defenders' forces and to ease the progress of the advancing troops.

Artistic portrayals of breakthrough with ladders display a number of clear differences between Egyptian and Assyrians reliefs. In the Egyptian reliefs, the ladders are set against the wall at an acute angle of 7°-23°, while in Assyrian reliefs the angle is wider (25°-46°). In the Egyptian reliefs, the combatants are shown climbing with both hands free to hold on to the ladder, their elliptical shields hanging on their backs.[95] A relief from Saqqarah (Fifth Dynasty) shows warriors climbing with axes hanging from their belts,[96] while in reliefs from the reign of Rameses II (13th century BC) and Rameses III (first half of the 12th century), warriors hold straight swords that do not overly obstruct their hold on the ladder with both hands.[97] In

century BC. Various opinions have been expressed as to its purposes, such as defense against battering rams (Yadin, *BASOR* 137 (1955), 23-32), defense against tunneling (Stager, *Biblical Archaeology Review* 17/2 (1991), 30) or for the protection of the wall foundations against erosion and to impede tunneling (Herzog, in *The Architecture of Ancient Israel from the Prehistoric to the Persian Periods,* 267). Without going into a discussion of these views, it is significant for our purposes that the glacis provided an excellent defense against attack ladders; cf. below, p. 88, n. 162 .

95 The hanging of shields on one's back was not intended as protection against enemy fire (as Yadin proposes, *Art of Warfare,* 229), since any arrows would come from the front, not from behind.

96 Yadin, *ibid.*, 147.

97 For the dating of this relief to the reign of Rameses II, see Yadin, *ibid.*, 228 (=*ANEP* 334). On the view that this relief is from the days of Merneptah and relates to

Assyrian reliefs from the 9th–7th centuries BC, attackers are shown climbing the ladders upright, with a (typically large) shield in their left hand and a spear in their right, without holding on. In reliefs from Ashurbanipal's reign, archers are among the attacking forces, and they too are shown climbing the ladders.[98]

If these are not merely artistic conventions that do not reflect actual military technique,[99] comparison between the Egyptian and the Assyrian portrayals reveals a development of assault techniques. Namely, during the Neo-Assyrian period soldiers were trained to run, without holding on, up the ladders (which were placed at a convenient angle for that purpose), thereby entering quickly into hand-to-hand combat with the defenders stationed on the wall (the archers shown in the Ashurbanipal reliefs were obviously intended to assume shooting positions at the top of the wall toward the city). This method of attack may also be mentioned in the Bible:

a) In the oracle of Joel 2:1ff., the appearance of a swarm of locusts and their advance is compared to that of an army. After approaching the city, "They rush like warriors, they scale a wall like fighters (כגבורים ירצון, כאנשי מלחמה יעלו חומה). Each keeps to his own track; their paths never cross; No

---

his war against Ashkelon, see Stager *EI* 18 [1985], 56*ff.); Yurco, *JARCE* 23 (1986), 207-215; 229 (=*ANEP* 333). For dating to the reign of Rameses III: Yadin, *ibid.*, 346 (=*ANEP* 344).

98 For the reliefs from the reign of Ashurnasirpal II: see Yadin, *ibid.*, 388; 392, no 4; from the reign of Tiglath-pileser III: Yadin, *ibid.*, 406; from the reign of Sargon: Yadin, *ibid.*, 424-425; from the reign of Ashurbanipal: Yadin, *ibid.*, 448, 449, 462 (=*ANEP* 10). An unusual depiction of climbing on ladders appears in the bronze reliefs on the gate of Balawat dated to the reign of Shalmaneser III: here climbing warriors are depicted with shields in their right hands, covering their fronts, and their left hands are free and apparently holding onto the ladder (in any event, there are not any attack weapons in their left hands); cf. Yadin, *ibid.*, 398 (=*ANEP* 359); *ANEP* 362, 365.

99 That the Assyrian soldiers did not hold onto the ladders during an assault is particularly evident in reliefs depicting precise, realistic technical detail (including activities related to siege and breakthrough, such as digging and crumbling walls with swords, separating bricks with pikes and holding onto the end of a battering ram with metal chains). See, e.g., Yadin, *ibid.*, 392-393 (on the reign of Ashurnasirpal II); 448, 462 (on the reign of Ashurbanipal). These strengthen the view that the method of climbing up on ladders depicted in these reliefs also reflects a technique that was customary at the time.

one jostles the other, each keeps to his own course. And should they fall through a loophole, they do not get hurt" (ובעד השלח יפלו לא יבצעו ; vv. 7–8)

b) 2 Sam. 22:30 // Ps. 18:30 reads: "With you, I can rush a barrier, with my God I scale a wall" (כי בכ(ה) אר(ו)ץ גדוד, (ו)באלהי אדלג שור). The phrases "they scale a wall like soldiers" and "with my God I scale a wall" clearly refer to the scaling of a wall during an assault. The poetic parallelism suggests that "they rush like warriors" and "with you I can rush a barrier" refer to similar issues.[100] Joel 2:8 may also refer to climbing a ladder: like soldiers charging the top of the wall, each following the line of advance set for him, and whoever hesitates or is injured by enemy fire falls or is removed so as not to hinder the others, so too the locusts ascend and approach in an organized manner. The term שלח in Joel 2:8 refers to an obstacle which the attackers need to pass; some think that this refers to a certain water obstacle, while others think it is another name for a wall.[101]

The technical nature of the assault ladders is quite clear, but the linguistic aspect is less clear: the word *simmiltu* (with or without the determinative GIŠ) is used in Akkadian to refer both to a ladder and to a flight of stairs,[102] and its specific semantic meaning is difficult to ascertain.

a) In the Akkadian omen literature[103] and in the queries to the gods,[104]

100 The parallel between the texts discussed here reinforces the assumption that the word אר[ו]ץ in 2 Sam. 22:30 // Ps. 18:30 is derived from רוץ, "to run" (rather than from רצץ, "to crush," as thought by various commentators). Accordingly, the word גדוד is interpreted here as an obstacle rather than as an army unit. Some scholars connect this word with גודא, Aramaic "wall" (possibly also city wall?); see Y. Kutscher, *Leshonenu* 27–28 (1963–64), 185–186 (and similarly in Löwenstamm-Blau, *Thesaurus of the Language of the Bible,* II, 195).

101 On the assumption that שלח is the outer wall (parallel to *ša/ulḫû*; see CAD Š/I 243–244), as opposed to *dūru* (=the inner wall); cf. Löwenstamm, *From Babylon to Canaan,* 19–21. For a relief depicting assault with ladders against the outer wall and the inner wall or walls of the city of Harhar in Media, see Yadin, *Art of Warfare,* 424.

102 See CAD S 273–275 *simmiltu.*

103 See Labat, *Textes littéraires de Suse,* 96 (No. IV) rev. 43: *nakru dūrī ina simmilāti* (GIŠ.KUN$_5$.MEŠ) *iṣabbat* (cf. also *ibid.*, 133, Text VI, l. 10); *ibid.*, rev. 44: *dūr nakri ina simmilāti aṣabbat*; *KAR* 446 rev. 6': *dūrka nakru ina simmilāti* (KUN$_4$) *iṣab[bat].*

104 SAA IV 43:7: *lu ina epēš kakkē qabli u tāḫā]zi lu [ina ni]ksi lu ina pilši simmilti*

the term ^{giš}*simmiltu* is mentioned as a method of conquering cities. The Assyrian version of the myth of Nergal and Ereshkigal speaks of ^{si}*simmilat šamāmī, "simmiltu* of heaven," in which the messengers of the gods ascend to heaven.[105] The clear resemblance between this ^{si}*simmilat šamāmi* and Jacob's ladder, placed upon the earth with its top reaching to the heavens, upon which angels (i.e., messengers) of God ascend and descend (Gen. 28:12), leads one to conclude that: (1) the biblical word סֻלָּם (a *hapax legomenon*) may be understood as a metathesis of *simmiltu*;[106] (2) Jacob's ladder may be seen as a flight of stairs (upon which one may simultaneously go up and down), as opposed to the ladders shown on the reliefs.[107]

b) There are those who derive the word סֻלָּם from the root סלל ("to heap up, to cast up, to pave a [high]way").[108] If this derivation is correct, than the words סֻלָּם and *simmiltu* are not related; on the other hand, the derivation סֻלָּם relates to the sequel of the description of locusts in Joel 2:8: "No one jostles another, each keeps to his own course" (גבר במסִלתו ילכון; מסלה is also derived from the root סלל).

c) As noted earlier, the conquest of a city by *simmiltu* occurs in the omen literature and in the queries to the gods, but not in the royal inscriptions. This fact, together with the large number of ladders appearing in Assyrian and Egyptian reliefs, led Landsberger to think that a different term, *kalbānātu,* is used in the royal inscriptions to indicate an assault ladder.[109] However, others opposed this assumption, particularly in view of the presence of the two terms together in the *tamītu* texts.[110] While it is agreed, particularly in light of etymological considerations (*nabalkutu,* "to cross over"), that

(GIŠ.I.BAL) *u nabalkatti*; cf. SAA IV 30:8, 44:9; IM 67692:54 *ina dimti ina āšibu ina ritti ina simmilti ina kalbānāti ina nikis dūri ina nabalkatti*; 81-2-4, 209:3 *[l]u ina simmilti lu ina kalb[ānāti…].*

105 See Gurney, *AnSt 10 (*1960), 122 (v 13'), 124 (v 42').

106 Cohen, *Biblical Hapax Legomena in the Light of Akkadian and Ugaritic,* 34 (and references therein).

107 In addition to the previous note, see also references in Houtman, *VT* 27 (1977), 337-338.

108 Houtman, *ibid*., 338-340.

109 See Heidel, *Sumer* 9 (1953), 179.

110 See above, n. 104.

nabalkattu indicates means of getting over a wall, evidently by a ladder,[111] CAD is noncommittal as to the precise definition of *kalbānātu* and settles for the general term, "siege engine."[112] Moreover, *AHw* questions whether the reading *kalbānātu* (as opposed to *labbānātu*, for example) is in fact the correct one.[113]

A comparison of the descriptions of breakthrough in the Sennacherib inscriptions (on the conquest of the cities of Judah)[114] and the inscriptions of Esarhaddon (on the conquest of Memphis)[115] offers some confirmation of Landsberger's view. According to this view, the sequence of the terms *pilšu, niksu* and *nabalkattu* in these passages may be explained as referring to the three methods of breakthrough: a) below the wall, b) through the wall and c) above it.

The problem becomes even more complex in light of the fact that, in addition to the two terms discussed here, *kalbānātu* and *nabalkattu,* a third term is added in the *tamītu* texts – namely, *simmiltu*, with which we began our discussion, a term which probably refers to a ladder. Does the existence of three terms, sometimes appearing alongside one another in the same sentence, require us to assume that they refer to three different methods of breaching the wall or going over it? The appearance of similar or parallel terms (such as *sunqu, ḫušaḫḫu* and *bubūtu*) alongside one another in the same text[116] indicates that, in the type of sources discussed here, extensive use is made of synonyms. Therefore, it is not unlikely that all three terms indicate assault ladders (possibly of different kinds).

Each of these approaches has its own advantages and problems, and it is difficult to decide among them definitively.

111 Compare CAD N/I 9 *nabalkattu* 2,a; *AHw* 694 *nabalkutu* II,a.

112 CAD K 67 tends to connect the term *kalbānātu* with *kalappu; cf.* Hebrew כילף (pick-axe).

113 See *AHw* 424 *kalbānātu*; 524 *labbānātu.*

114 OIP II 33:23: *ina šukbus aramme u qitrub šupê mitḫuṣ zūk šēpē pilši niksi u kalbannāte alme akšud,* "By treading down the ramparts and bringing battering rams close, infantry attack, tunneling, breaching, and *kalbannāte* I waged war and I defeated them."

115 Borger *Asarh.* 99:42: *ina pilši niksi nabalkati alme akšud,* "By means of tunnels, battering rams and *nabalkati* I attacked and I conquered."

116 See, e.g., IM 67692:59; SAA IV 29 4'.

2. Tunnels

Archaeological evidence of digging, either through the wall or beneath it, as a method of conquering a city, is found from the 9th century BC onwards.

In the excavations at Kouklia (Palaepaphos), three tunnels (*Stollen*) were discovered in Area A, hewn (in limestone) or dug (in the earth) beneath the wall. They were 15 to 20 meters in length, 1 to 2 meters wide, and one of them was as high as 2 meters.[117] The three tunnels terminate inside the city, about three meters beyond the wall. The excavator, F. G. Maier, thinks that they were dug by the defenders of Paphos, who had planned to fill them with flammable material and set fire to them in order to undermine the siege ramp which the Persians had built to attack the city. I doubt this view and suggest that the tunnels were dug by the attackers rather than by the defenders.[118] The width of the tunnels was clearly insufficient to cause the collapse of the wall (which was 5 meters wide),[119] and therefore, it seems that they were intended for smuggling combatants into the city.

When the Middle Bronze Age glacis in the west of Tel Lachish was excavated, a tunnel was also discovered that had been dug through the debris that constitutes the glacis. The width of the tunnel in the excavated section was about 1.20 meters.[120] The excavator, J. L. Starkey, thought that the tunnel was dug during a struggle over the city in the time of the Eighteenth Dynasty in Egypt (15th century BC).[121] Yeivin, argues that there is no evidence for the use of tunnels for penetration beneath fortifications prior to the time of Ashurnasirpal II, and attributes this tunnel to the siege on Lachish and its conquest by Sennacherib in 701 BC.[122] However, this view must be examined in light of the linguistic and literary considerations.

117 See the bibliography on the tunnels of Kouklia, above, pp. 28–30, and see there, cross-section B-B'. Diagrams of the other tunnels were not detailed in the publications available to me.

118 See the discussion above, pp. 28–30.

119 On collapsing sections of the wall by tunnels, see Arrian, 2.27.4 (Alexander's war against Gaza).

120 See Tufnell et al., *Lachish,* IV, Plates Pl. 6:2.

121 Starkey, *PEF QSt* 1934, 169.

122 Yeivin, *EI* 1 (1951), 29–31.

Depictions of the undermining of city walls and of tunneling beneath them appear in reliefs from the palace of Ashurnasirpal ("the North-Western Palace") at Calah, in which siege warfare is depicted:[123] in relief BM 124554 (in the bottom-center of the relief), two Assyrian soldiers are shown in a tunnel at the bottom of the wall (one handing an object of unclear nature to the other), while two other soldiers (on the left of the relief) are seen carrying a pike and a spear (or a metal object resembling a spear), and removing bricks from the wall. The latter two stand exposed outside the wall.[124] In relief BM 124552, a soldier is shown on the bottom right-hand corner, bending down or lying in a tunnel while digging with a dagger held in his right hand. Digging through or undermining an exposed wall (in several places simultaneously) with swords or daggers also appear in the depictions of the battles on the Elamite city of Hamanu (in relief BM 124931) and on Thebes in Egypt (in relief BM 124926) from the palace of Ashurbanipal ("the North-Western Palace") in Nineveh.[125]

Thus, the reliefs depict two forms of breaching: 1) digging through the city wall above ground; 2) tunneling. Digging through the wall on the surface was relatively rapid and convenient; but in order to reach the wall, the soldiers needed protection and cover, as they did during the process of digging. By contrast, burrowing in through tunnels was more secure until the stage of bursting out above ground in enemy territory (the danger

123 Yadin, *Art of Warfare,* 388, nos. 4-5 (BM 124552, 124554); 392-393 (enlarged photograph). For enlarged photographs of these reliefs, see Barnett and Lorenzini, *Assyrian Sculpture in the British Museum,* Pls. 27-29.

124 These soldiers seem to be standing next to a water canal. It may be a moat, and the battering was done by hand because the moat prevented the approach of battering rams.

125 Yadin, *ibid*., 448 (Hamanu), 462-463 (Thebes). The soldiers breaching the wall hold a dagger in one hand and protect their heads with shields held in the other hand (some shields display embedded arrows that were shot from the wall). Compare *ANEP* 362: the Balawat gate (from the reign of Shalmaneser III) depicts a soldier digging the wall of Dabigu (in Bit-Adini), holding a small shield in his left hand and a spear in his right hand. For textual evidence of this form of breakthrough, compare IM 67692:58 *ina patār pālši* – that is, by the digging dagger; Ezek. 26:9: "He will direct the shock of his battering rams against your walls, and with his axes he will break down your towers." I wish to thank Prof. Moshe Greenberg for calling my attention to this verse.

involved in this stage will be elaborated below), but this method was much slower.

Herodotus mentions tunneling beneath the walls of fortified cities in connection with the conquest of cities by the Persian army during the reign of Darius I (Barca, 4.200; Samos, 5.115; Miletus 6.18).[126] Similarly, it is recorded that Alexander the Great penetrated the walls of Gaza through tunnels.[127]

The Semitic terms used to signify the various forms of boring pose a problem in light of the presence of two terms in the Assyrian documents, *niksu* and *pilšu*, that frequently appear alongside one another,[128] and in light of the differing opinions as to their precise meaning.[129] From the expression *nikis dūri,* it follows that *niksu* refers to breaching the wall rather than tunneling beneath it. The Hebrew semantic equivalent is therefore פרץ (compare also the Syriac תורעתא, used in the Peshitta to translate that word). In biblical law and in the Code of Hammurabi, which are concerned with the punishment of a thief discovered during a break-in, we find the terms מחתרת (Exod. 22:1) and *pilšu* (CH §21); compare also the Peshitta to Exodus 22:1 (מחתרת): פולשתא. However, from these facts and from the occurrences of the verbs חתר and *palāšu*, it is impossible to determine whether the terms refer to digging beneath the wall alone, or whether they also refer to penetration of the wall above the surface.[130] Therefore, the precise semantic relationship between the terms *niksu* and *pilšu* cannot be determined by linguistic means.

The conquest of a city by a *pilšu* appears in Akkadian documents from the Old Babylonian Period onward, e.g., many omens mention the conquest

126 Compare also Polyaenus, 7.11.5, on the long tunnel dug by the army of Darius I during their siege of the city Chalcedon.

127 Arrian 2.27.4-5; compare Curtius Rufus 4.6.21-23.

128 In the Akkadian lexicons, there is no real distinction drawn between the two terms; CAD translates *pilšu* – tunnel; *niksu* – breach, tunnel. In the 1990 edition of the queries to the Sungod (SAA IV), *pilšu* – breach and *niksu* – tunnel; see *AHw: pilšu* – Einbruchsstelle, Bresche; *niksu* – Durchschneiden, Bresche.

129 OIP II 33 iii 23; Borger, *Asarh.* 99 rev. 42; SAA IV 31:7, 43:7.

130 Cf. Ezek. 8:8; 12:5, 7; Job 24:16; CAD P 58-62: *palāšu,* "to pierce, bore, to break through, break into."

of the city of Apishal by Naram-Sin (23rd century BC) by a *pilšu.*[131] The same method is mentioned in 18th century BC Babylonian documents (in the inscription of King Dadusha of Eshnunna, viii 6, the conquest of Qabra; ARM I 135:8-10, the conquest of uru*Qirḫum*). The term appears again in Assyrian royal inscriptions from the 9th -7th centuries BC,[132] as well as in queries to the Sungod and in *tamītu.*[133] Their mention in these inscriptions is not indicative in itself of which of the two methods of breakthrough was used. Information concerning the duration of battles in which cities were conquered by *pilšu* may indicate something regarding this question: Qabra was conquered on the tenth day, and Qirhum on the eighth day.[134] One must determine whether it was possible to dig tunnels within those time periods, or whether the unavoidable conclusion must be that the cities were breached from aboveground.

The critical stage that determined whether the digging of a tunnel was successful or whether all the effort invested in it was in vain was the moment when the diggers and the attackers jumped out to the surface at the far end of the tunnel within the city. Due to the great vulnerability of the warriors in the tunnel (see Aeneas, below) and the ease with which the city's defenders could block off the end of the tunnel once they became aware of its location, it was of vital importance that the conclusion of the digging be performed as quickly and as stealthily as possible. Nevertheless, although

131 These documents are much later than the reign of Naram-Sin. Some scholars have questioned whether references in the omens (and in other genres based on them) concerning the conquest of Apishal by *pilšu* in fact indicate a historical event. See: Reiner, in *Anatolian Studies Presented to H. G. Güterbock,* 257-261; Cooper, in *Death in Mesopotamia,* 99-105. For details of the inscriptions in which this event is mentioned and a discussion thereof – while rejecting the assumption that they are nothing but a wordplay on the consonants in the name of the city, *pšl* and *plš* (French: *saper*) – see Glassner, *RA* 77 (1983), 3-10.

132 RIMA 2 216:53, 220:111 (Ashurnasirpal II); RIMA 3 191:15' (Shamshi-Adad V); OIP II 33 iii 23; K 6205 + BM 82-3-23, 131:8 (Naʾaman, *BASOR* 214 [1974], 26) Judah; OIP II 83:45, Babylon (Sennacherib); Borger, *Asarh.* 99 rev. 42 (Esarhaddon).

133 SAA IV 31:7, 43:7, 44:9, 102:4'.

134 Dadusha Inscription, viii. 5-9. (It should be noted, however, that according to l. 6, a siege rampart was erected. Accordingly, does *pilšu* in this context refer to the breach made by a battering ram?); ARM I 135:11-13.

the besieged inhabitants knew that tunnels were being dug underneath the walls of their city (in light of the frequency with which this method was used), it is worth noting that it was often difficult to thwart such a plan because it required constant vigilance on their part, as they did not know when or where the attackers would penetrate from underground.[135] Moreover, in light of the danger and vulnerability involved in this method, it seems reasonable to assume that this was not the only method of attack used and that penetration from the tunnel was carried out while diversionary activity involving attack on the city took place elsewhere.

The moat (Akkadian: *ḫarīṣu*/*ḫirīṣu*, *ḫirītu*) was a reliable method against tunneling. Once the end of the tunnel was discovered in the wall of the moat, it was possible to harm the attackers within by filling the tunnel with smoke or sending hornets and bees inside (Aeneas, Ch. 37); one could, of course, also shoot from the wall at the soldiers seen emerging from the open end of the tunnel. To better prevent the attackers from penetrating the city by tunnel without being discovered, it was preferable that the moat reach a lower layer, impenetrable to the diggers. Indeed, in Assyrian and Babylonian royal inscriptions, we find the following data:

Tukulti-Ninurta I (13th century BC) surrounded the city wall of Assur with a moat hewn into the rock to a depth of 20 *mušaru* (possibly 20 cubits),[136] to groundwater level.[137] Nur-Adad of Nisibis (late 10th–early 9th century BC) surrounded the walls of his city with a nine-cubit-wide moat which also descended to groundwater level.[138] Merodach-baladan fortified the city of Dur-Yakin against King Sargon of Assyria and surrounded its wall with a moat, 200 cubits wide and 1.5 NINDA (=21 cubits) deep,[139] thus reaching groundwater.[140] The description of the fortifications of Ashdod during the

135 Herodotus 4.200, in writing about the siege of Barca, relates the discovery of a Persian tunnel, after it had already passed beneath the wall into the city, by placing a brazen shield on the ground in various places. When the shield was placed above the digging point, the noises created an echo, while at the other points the shield was silent.

136 On *mušaru* as a unit of length used to designate depth in excavation projects, see Powell, *RLA 7:*492

137 RIMA 1 267:6–8.

138 RIMA 2 151:64–66 (Adad-nirari II).

139 On the ideogram NINDA and its length (14 cubits), see Powell, *RLA* 7:471.

140 Lie, *Sar.,* 58–60:405–406 (the text of these lines, including the numbers and the units

preparations for rebellion against Sargon reports that it was surrounded by a moat, 20[?] cubits deep, which reached groundwater[141]; during the reign of Nebuchadnezzar II, the eastern wall of Babylon was fortified and a moat was dug that reached groundwater, whose walls were calked with bitumen and with burnt bricks.[142] In places that lacked an impenetrable layer, the attackers could dig deeper and tunnel beneath the moat. This was probably the situation described in the Zakkur Inscription (*KAI* 202 A), line 10: "They put up a siege ramp higher than the wall of Hadrach, and dug a trench deeper than its moat."[143] The enemies, King Ben-hadad of Aram and his allies, had made significant progress in their siege of Hadrach: they had already attained sufficient height to dominate the view of the city and attain superior firepower over the walls of the city and had already dug beneath the moat.[144] Their armies were thus ready to attack the city. At this decisive stage, "I lifted up my hands to Baalshamayn, and Baalshamayn answered me, and Baalshamayn [spoke] to me by seers and messengers(?)" (ביד חזין וביד עדדן; ll.11ff.).[145]

---

of measure, is damaged. Its restoration has been based on parallels from Hall V in Sargon's palace at Khorsabad). On the significance of the operation in terms of the quantities of soil excavated and the labor invested therein, see Powell, *JCS* 34 (1982), 59–61.

141 Winckler, *Sar*. 188:23–25. See the reconstruction of these lines in Na'aman, *Shnaton* 2 (1977), 171.

142 *Nebuch*. 1 ii 17–18; cf. Langdon, *Die neubabylonischen Königsinschriften,* 74.

143 והרמו שר מן שר חזרך והעמקו חרץ מן חר[צה]. Although the meaning of the text is clear, it should be noted that the usual sense of the terms שר and חרץ is not "siege ramp" and "tunnel," but rather "wall" and "moat." The use of these terms here, as parallels of שר and חרץ, is attributed to the poetic style. On the nature of the Zakkur Inscription as a *Danklied,* see Greenfield, *ʿAl Kanfei Yonah, 75–92*.

144 On the integration of these two techniques in battles for cities, compare: Ashurnasirpal inscription, RIMA 2 216 iii 53: *ina pilše nāpili ṣāpīti āla aktašad*, "I conquered the city (Kaprabu) by tunnels, battering rams (and) siege towers"; *ibid*., 220 iii 111: *ina pilši* ^{giš}*ṣāpīte u nēpēšē āla akšud*, "I conquered the city (Udu) by tunnels, siege towers and *nēpešē*" (on the meaning of *nēpēšē*, see below, pp. 100–102). Curtius Rufus on Alexander's battle against Gaza (4.6.21–23): the artificial hill that was heaped up (against) the wall reached the height of the wall itself; the besiegers placed towers on it and shot from them into the city, which was exposed to their fire.

145 On the king's appeal to the god(s) at the decisive stage of the battle for the city, see discussion below, pp. 152–153.

3. Breaching with Battering Rams

Breaching the wall is a method of penetration which does not employ the element of surprise and, compared to other methods of breakthrough, requires considerable manpower, specialized equipment (battering rams)[146] and an extended period of time.

There are written accounts from the middle of the 3rd millennium BC onward of breaching walls using battering rams (and siege ramps that were required to elevate them and bring them close to the walls, which were placed upon hills or mounds): the Ebla documents mention battering rams and other breaching devices (GIŠ.MÁ.NE=*mekû*),[147] while Old Babylonian inscriptions from Mari and from Boghazköy and later Neo-Assyrian inscriptions mention battering rams and siege ramps separately or in conjunction.[148] There is no evidence in Egyptian sources for the use of siege ramps and battering rams; consequently these would seem to have been limited to the Mesopotamian-Anatolian region.[149]

The battering ram consisted of a wooden pole with a metal point that was thrust against the wall, preferably in the crevices between the bricks or stones, to destabilize it.[150] Over the course of nearly a thousand years, from the 18th century BC onwards, inscriptions mention the construction of siege ramps alongside the operation of the battering rams themselves. This suggests that the basic method of operation did not change. Nevertheless,

146 From the portrayals in the Assyrian reliefs, it follows that breaching the walls was sometimes carried out manually, with pikes and swords; see, e.g., Yadin, *Art of Warfare,* 392, 448, 462; and see above, p. 77, n. 125. The capacity of battering rams was much greater, of course. On the manner of operation of the battering ram in piercing the walls, compare also the Sumerian expression, GIŠ.GU$_4$.SI.AŠ, that is, "a one-horned bull"; Hebrew: כר (Ezek. 4:2; 21:27) = "ram"; and in Western languages, κριός, *aries,* and so on.

147 See Steinkeller, *NABU* 1987, no 27.

148 See CAD A/II 428–429 *āšibu.*

149 Note the term "a battering ram of Hurrian type" (in the story of the siege of Urshu), KBo 1 11 obv. 15' (Güterbock, *ZA* 44 [1938]), 116; Beckman, *JCS* 47 [1995], 24).

150 The verbs used in conjunction with *āšibu,* that may be indicative of its purpose, are *abātu, maḫāṣu, qerēbu* and *purruru.* It seems that the heads of the rams differed in size. The letter ARM XXI 71bis:19–21 refers to a point that weighed one mina (= about ½ kilogram).

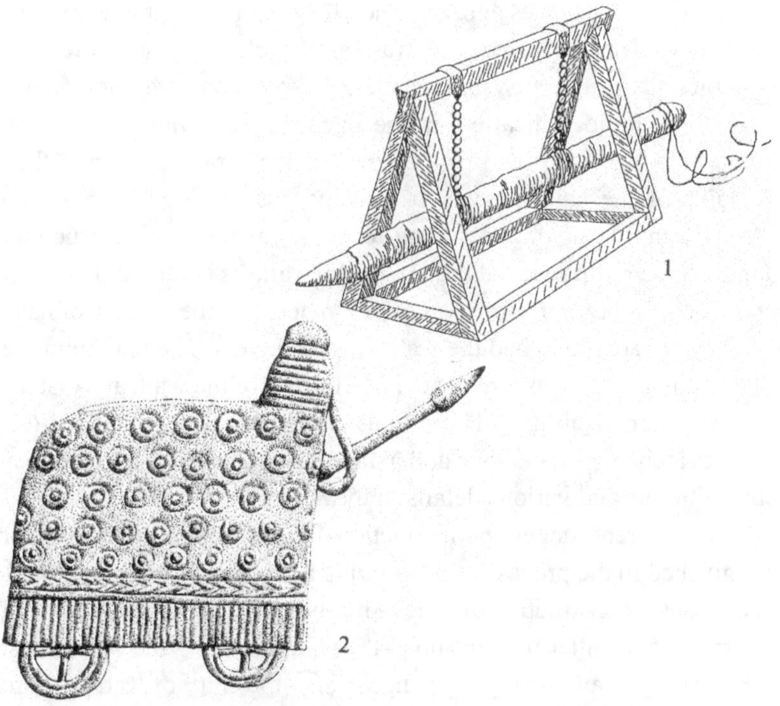

Fig. 6: Battering rams. (1) Reconstruction of Scurlock, *SAAB* 3 (1989), 130.
(2) From the time of Tiglath-pileser III
From Yadin, "Weapons, Weaponry," *Enc. Miqr. 5: 971-972*

examination of Assyrian reliefs from the 9th and 8th centuries BC reveals that battering rams were improved over a brief period, in a way that increased their mobility and possibly the method of their operation as well.[151] It seems reasonable to conclude, therefore, that these weapons underwent substantial improvement and changes during their long military use.

151 On the development of battering rams between the reigns of Ashurnasirpal II and Sennacherib (battering rams do not appear in the Ashurbanipal reliefs), see Yadin, *Art of Warfare,* 314-316; *idem.*, "Weapons and Weaponry [vii]," *Enc. Miqr* 5.967-970 and the table in cols. 971-972.

Up until a few decades ago, our knowledge of the technique involved in breaching walls by battering rams was based exclusively on written sources and depictions in the Assyrian plastic arts.[152] A concrete component was provided by the identification, in the Lachish excavations, of portions of an external siege ramp and of a counter defense-rampart inside the wall – remains of the battle in which the city was defeated by Sennacherib of Assyria in 701 BC.[153] According to the excavators' reconstruction, the external siege ramp reached the top of the wall, its central axis was 50–60 meters long, its bottom width was 50–75 meters and the breadth of its upper extremity, where it touched the wall, was 25 meters. The maximum height of the counter-ramp was 3 meters above the top of the wall, at a distance of some 10 meters from it.[154] These finds, combined with extant written and artistic evidence, promote our understanding of the significance of several technical terms and various details in the Assyrian reliefs.[155]

Three different stages, both functionally and chronologically, can be distinguished in the process of penetrating a city through its outer wall: (1) preparations – construction of a siege ramp reaching the wall, (2) ramming – transfer of the battering rams up to the wall, ramming the wall, smashing and breaching it, and finally (3) penetration – the entry of warriors through the breach in the wall and their deployment within the city. These stages are outlined below, particularly with reference to the deployment of forces and the activities and circumstances characterizing each stage.

(a) Preparations: Construction of the Siege Ramp

Gathering the large quantities of materiel needed for the siege ramps within a short period of time required extensive (albeit unskilled) manpower and efficient organization.[156] Since the porters were numerous and exposed

152 See Yadin, *Art of Warfare,* esp. 16–24, 313–462; de Vaux, *Ancient Israel: Its Life and Institutions,* 236–238.

153 For the final report on these ramparts, see Ussishkin, *The Renewed Archaeological Excavations at Lachish (1973–1994)*, 695–764.

154 Ussishkin, *ibid.*, 716.

155 See Ephʿal, *TA* 11 (1984), 60–70 (substantial parts of this article appear in the present discussion of battering rams); Ussishkin, *ibid.*, 716ff.

156 Estimated data on the effort needed in the construction of the Assyrian siege ramp

to enemy fire – as it is impossible to hold a shield while carrying stones and baskets and it is hard to move while the body is protected by armor – it seems likely that night was the preferred time to build the ramp, in order to significantly reduce the danger of fire from the city defenders. Nevertheless, the time pressure would have made it impossible to completely

---

in Lachish illustrates the magnitude of the undertaking: the weight of material in the rampart was 19,000 tons (on the dimensions of the rampart, see Ussishkin [above, n. 153], 716–717.) The labor output of a porter per hour: 3 rounds of 15 kg. each, i.e., 45 kg., over a distance of 300 m. Thus, 500 porters could carry 22.5 metric tons per hour, so the transport of 19,000 tons required 844 man-hours (19,000/22.5) – 35 days (i.e., 844/24) of continuous labor in shifts of 500 porters each (on the construction of a ramp in 70 days of continuous labor day and night, see Thucydides, 2.75: the siege of the Lacedaemonians upon Plataea, 429 BC). Assuming ten-hour work shifts, according to the specifications above, the transportation of construction material alone required 1,200 porters per day. The work force must have also included the laborers required to gather stones, fill baskets and other work.

These estimated numbers raise the question whether the laborers engaged in laying the ramparts at Lachish were Assyrian soldiers or captives from Judah or Philistia, who were taken prisoner during Sennacherib's campaign. The employment of captives had an additional advantage: the defenders were unlikely or at least reluctant to shoot at the laborers, their captive brethren, to dissuade the besiegers from approaching the city walls. On the use of prisoners of war as protection against the fire of their brethren, see Frontinus, *Stratagems* 1.4.1–2.

The letters from Mari and Shemshara tell of the conscription of thousands and even tens of thousands of military men for operations that included a siege (ARM I 42; Shemshara, No. 64). The large numbers are surprising considering the assumption that it was impossible to control such a large number of combatants in field battle. Yet, the credibility of the information is increased by the fact that the source are letters rather than royal inscriptions or reports from later historians (e.g., Diodorus 17.40.5, who relates that Alexander the Great employed tens of thousands of people, including all the inhabitants of the environs, in the construction of the causeway which he laid down to connect Tyre to the mainland). These numbers (which seem extreme) may be explained if they are taken to refer to the number of people who were engaged in the extensive earthworks required in order to operate battering rams in a short period of time. This is exemplified in the inscription of King Dadusha of Eshnunna (ll. 90–94 = Spalte viii 5–9 in Ismail, *BaM* 34, 2003, 144) which notes that the city of Qabra was conquered on the tenth day of its siege, and after ramp and tunnels had been employed against it. On the size of the armies mentioned in the Mari documents in various contexts, not necessarily related to siege, see Abrahami, in *La circulation des biens, des personnes et des idées dans le Proche-Orient ancien,* 157–166.

Fig. 7: Porters supervised by Assyrian soldiers, carrying earth, bricks and stones and gathering them in a heap.
Relief from Sennacherib's palace at Nineveh
From Paterson, *Assyrian Sculptures: Palace of Sinacherib*, Pl. 34-35

forego daytime work, which required more suppressive fire. The closer the construction work on the rampart was to the wall, the necessity for suppressive fire increased as well.[157]

Excavations at Lachish revealed that the Assyrian rampart was made mostly from stones. Tens of thousands of stones, heaped in no particular order, one on top of the other, served as a solid base with high weight-bearing capacity that could be built quickly – unlike earthen ramps built without stones, whose construction and solidifying required extended labor, possibly including mixing with water. It should be noted that siege ramps were not intended to provide stability for permanent buildings, as was the function of ramps that served as foundations for walls or buildings. Siege ramps were only intended to carry siege devices and soldiers for a very limited period of time. For this purpose, a base of piled rocks was sufficient. The rocks were covered by a layer, approximately one meter thick, of pressed earth or mortar (a mixture of sand and lime or other bonding material), which provided a surface for fighting and for the transportation of machines and men.

Akkadian has two technically distinct terms to indicate activities involved in the construction of a siege rampart: (a) *ep(e)rī šapāku* (lit., "piling up (loose) earth"), the gathering of soil (and stone) and their piling up (while still soft and malleable);[158] (b) *arammu šukbusu*, the compression of these materials through trampling by foot. The parallel Hebrew term, כבש, is semantically close to דרך and סלל.[159] Assyrian scribes generally

157 Details of the technique of suppressive fire, used to cover those who approached the wall at the various stages of warfare, may be derived from the incorporation of the data in the depiction of the battle of Lachish (in the relief found in the palace of Sennacherib in Nineveh) and the discovery of the Assyrian siege ramp at Tel Lachish; see Ephᶜal (above, n. 155), 338–340.

158 See: CAD E 187, esp. *eperu* 2; Š/II 413–414 *šapāku* 1,2'. In the Aramaic Targum, the biblical expression שפך סוללה is rendered as צבר מליתא; compare the description of the advance of the Babylonian army, which did not face any obstacles: "They laugh at every fortress, for they heap up earth (ויצבר עפר) and take it" (Hab. 1:10). צבר עפר is used in this *hapax legomenon* to indicate the entire process of laying the ramp, and in essence the entire process of breaching the wall.

159 See CAD A/II 227–228, *arammu*; K 11, *kabāsu* 7c (all the quotations are Neo-Assyrian).

used one of these expressions in writing about siege activities,[160] but the two are combined in some works of a more literary-rhetorical nature.[161] The biblical expression שפך סוללה is thus a complex idiom that combines components of both Akkadian technical terms: שפך is parallel to the verb found in *ep(e)rī šapāku* and סוללה indicates the finished product of the verb found in *arammu šukbusu.*[162]

Beams and branches were sunk into the upper layer of a siege ramp to enhance its stability, in order to prevent the battering rams from slipping

160 See, e.g.: Heidel, *Sumer* 9 (1953), 150:38: *ina šipik eperī u qurrub šupê mitḫuṣ zūk šēpē iṣbatū āla,* "They (the soldiers of Sennacherib) besieged the city by pouring out earth, bringing close rams, and attack of foot soldiers" (On the reading š*ipik* here, compare Heidel's note, *ibid.*, 182); *ibid.*, 136:61–64: *ina šukbus arammē u qitrub šupë mitḫuṣ zūk šēpē pilšē niksī u kalbannāte alme akšud,* "I besieged and captured (the cities of King Hezekiah of Judah) by stamping earth (=constructing of siege ramps), by bringing up battering rams, by infantry attack, by tunnels, breaches, and by *kalbannāte* (a breakthrough device whose exact nature is not known; see above, pp. 74–75).

161 Compare Esarhaddon's Letter to the God, Borger, *Asarh.* 104 i 37: *arammu ina šipik eprī iṣṣē u abnē marṣiš pašqiš [ušakbis],* "[I stamped down] a rampart by pouring down earth, trees and stones with great effort and labor." For the reconstruction of the verb in the final word, compare *ibid.*, ii 2: *arammu ša eli* [uru]*Uppume āl šarrūtišu ušakbisu.*

162 Examination of the siege ramp leads us to wonder about the purpose of the glacis, which formed an integral part of city walls in Palestine in the Iron Age II; it is clear that by including the glacis within the siege ramps the attackers had a ready-made, solid filling and were saved a great deal of labor (see above, pp. 70–71). If Yadin is correct in his view (*BASOR* 137 [1955], 23–32) that the glacis served as a protective-operative device, that is, that it was intended to prevent penetration into the city by sapping the bottom of the walls, there is room for speculation regarding the best combination of the opposing interests involved in protecting the various parts of the wall: preventing the enemy from breaching the bottom of the wall on the one hand and, on the other hand, forcing the enemy to build a siege ramp under fire, while investing the maximum amount of materials and labor, in order to attack the upper part of the wall. Such a combination can be obtained by building a glacis with a steeply-angled incline, which could not be climbed by warriors or military devices without modification of the slope by the construction of a siege ramp. Alternatively, if the glacis was built first and foremost in response to static engineering needs and not due to tactical needs – that is, to prevent erosion at the base of the walls (on this approach, see, e.g., Herzog, in *The Architecture of Ancient Israel,* 267) – it must be concluded that the glacis was so necessary that it was built *despite* the awareness that an enemy might utilize it for the construction of a siege ramp against the city wall.

and sinking, and particularly in order to anchor them (this was required to absorb the vibrations from the shaking of the pole and its blows against the wall). The description of Esarhaddon's battle against Shubria recounts that the ramp prepared by the Assyrian army against the city of Uppume contained trees, which the besieged tried to set on fire.[163] This same phenomenon is evident in the description of the opening stage of the battle over Jerusalem: "For thus said the Lord of Hosts: Hew down her trees, and raise a siege ramp against Jerusalem…" (Jer. 6:6).

As the ramp drew close to the wall of the city, the vital stage of preparation for the advance of the breaching devices was completed.[164] The conclusion of this stage frequently heralded the imminent fall of the city under attack. It was at this stage that Jeremiah was commanded by the word of God to perform the symbolic act of buying Hanamel's field in Anathoth. The prophet's astonishment at this command and its timing is echoed in his words: "Behold, the siege ramps have come up to the city to take it; and the city, because of sword and famine and pestilence, is at the mercy of the

163 Borger, *ibid.*, 104 i 37. Further on, the text relates that the besieged poured naphtha (*napṭu*) on the ramp at night and set fire to it; however, thanks to a miracle performed by the god Marduk, the direction of the wind changed and the fire caught hold in the wall of the city and burned it down to ash (*ditalliš*): *ibid.*, ii 3-7; cf. below, p. 107, n. 213. This does not seem to have been an attempt to set fire to the battering rams of the Assyrian army, as steps were taken to protect these from fire at all hours of the day (according to the relief of the Lachish battle, a soldier was stationed on each battering ram whose job was to pour water on the ram in protection from the numerous torches thrown at it from the city wall; see Yadin, *Art of Warfare,* 314-315, 434-435), and they were apparently removed from the city walls at night; see below, p. 91. It is more likely that the besieged attempted to set fire to the wood that was embedded in the outer layer of the ramp to hinder the advance and operation of the battering rams. The wooden sections of the wall that were burnt were probably the balconies or projections at the top of the towers, which were constructed on long wooden beams that were laid down in crisscross fashion and protruded outwards, thus serving as convenient firing positions for the defenders (for a bronze model of such a balcony, see Boardman, *The Cambridge Ancient History, Plates to Vol. III,* fig. 92). Another possibility is that fire was set to parts of the wall in which wood was embedded; on a special device, which enabled the setting fire to such a section of the wall, see Thucydides 4.100. On setting fire to city walls, cf. Amos 1:7, 10, 14; Jer. 49:27.

164 Compare ARM I 4:9-16: "until the earthen ra[mparts] did not reach the height of the top (of the wall of) the city (*adi mūlê qaqqad ālim*), he had not captured the city."

Chaldeans. What you threatened has come to pass – as you see.[165] Yet you, Lord God, said to me: Buy the land for money and call witnesses – when the city is at the mercy of the Chaldeans!" (Jer. 32:24-25).

(b) Ramming: Transfer and Employment of the Battering Rams and Breaching the Wall

The durability of walls under attack by battering rams seems to be a function of their thickness rather than of the hardness of the building materials. The metal point of the ram, forcibly thrust against the dry bricks – the main building material of ancient walls – easily disintegrated them (consequently, the destruction of the walls during the stage of ramming did not last more than a few days). Nevertheless, the operation of the ram itself was not an easy task; Yadin argues that the ram, the metal-pointed pole which served as the active part of the battering ram, was suspended by ropes from the front of the roof of the support frame that held it, and was swung forward like a pendulum gathering momentum against the wall facing it.[166] Accordingly, the ram's angle of operation depended, as in a crane, on the length ratio between the parts of the suspended pole, which dictated the width of the section on which the ram could operate when the vehicle was fixed in its place. If the depiction of battering rams in the relief of the battle of Lachish is accurate, with more than half of the ram projecting outside of the vehicle, then its angle of operation was quite large (it is technically possible to balance the ram, suspended like a crane with unequal arms, by increasing the weight of the shorter arm). However, a device of this type would clearly be a burden to the operators within the vehicle. By contrast, if the inner arm was lengthened for the operators' convenience, the ram's angle of operations was shorter and consequently

165 The words "as you see" are missing in the Septuagint.

166 Yadin, *Art of Warfare,* 314; *idem*., "Weapons, Weaponry," *Enc. Miqr*. 5.967. In fig. 6 above, p. 83, one can see the wooden pole with its metal point, suspended by ropes tied to two points on the support frame. This type of hanging has a striking disadvantage: it does not allow the head of the ram to be raised, lowered or traversed. The basis for our current discussion is the assumption that the ram was hung from one point alone, thus facilitating movement of the ram as needed, although this method required more effort by the operators to stabilize the pole.

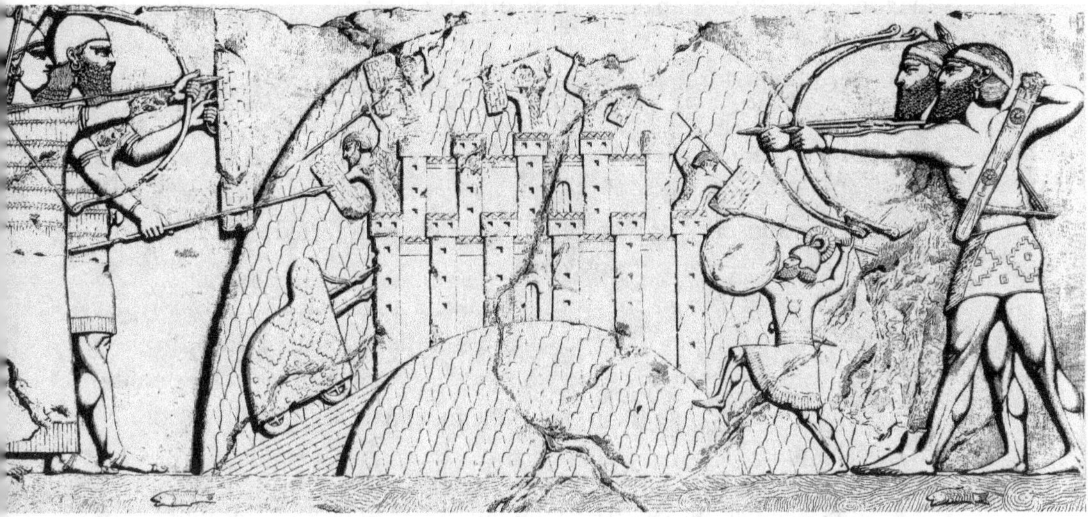

Fig. 8: An Assyrian attack on a city. A battering ram smashes a wall
From Botta & Flandin, *Monument de Ninive*, I/II, Pl. 77

the need increased to reposition the vehicle in order to enlarge the breach in the wall.

The operation of the battering ram required few people, but its defense, from the moment it was transported to the siege ramp until the wall was breached, whether in use or idle, required the continuous employment of dozens of archers at every watch, as the defenders shot at it with anything they could get their hands on, from stones and arrows to burning torches.[167] It seems likely that under such circumstances, the attackers would want the rams to operate vigorously and efficiently. The ideal time for their operation was during the daylight hours, when the warriors standing on

167 The relief of the battle of Lachish, depicts broken ladders and burning parts of chariots or wagons that were thrown off the wall. See Ussishkin, *The Conquest of Lachish by Sennacherib,* Segments III, IV; and pp. 101, 105. A particularly instructive passage on this issue appears in Judg. 9:53–54 concerning the circumstances of Abimelech's death in Thebez: he was killed by an upper millstone which was thrown at him by a woman upon the wall; cf. also 2 Sam. 11:20–21.

the fire base on the ramp could shoot effectively at the wall. Doubtless, the defenders did not cease their efforts during the hours of darkness, when the cover supplied from the siege ramps was less effective (it appears that the operation of the rams was also very difficult in the dark). The question whether the vehicles were taken off the ramparts every evening and returned in the morning (following a standard procedure of defending the rams during transport) or whether the rams were protected on the ramparts during the night remains unanswered.

Against what part of the wall was the ram directed? Ussishkin has observed that the relief of Sennacherib's battle against Lachish displays the rams battering the top of the wall and destroying the balconies on top of the towers rather than focusing on the main part of the wall. According to Ussishkin:

> It seems that the Assyrians did not intend to destroy the walls; how could their battering rams have forced a brick wall ca. 15 m. thick?[168] Instead they directed their attack at the balconies on the top of the wall and at the warriors manning them... Once the balconies crowning the wall have been destroyed and their defenders liquidated, nothing prevented the attackers from surmounting the wall and penetrating the city.[169]

At first glance, Ussishkin's theory seems convincing, but one cannot ignore the fact that, in many reliefs from the 9th and 8th centuries BC, the rams are clearly shown battering the *body* of the towers, not their tops.[170] An additional point is evident in the relief from the palace of Ashurnasirpal

168 This question is raised even more forcefully regarding the scenes in which Assyrian soldiers demolished the walls of cities with swords or pikes which they held in their hands.

169 Ussishkin, *The Renewed Archaeological Excavations at Lachish (1973–1994)*, 740–741. For additional examples from other reliefs of similar depiction of the operation of the rams, see Yadin, *Art of Warfare,* 408, 413 (from the reign of Tiglath-pileser III); 423 (the right hand ram); 425 (the left hand ram – from the reign of Sargon).

170 See, e.g., Yadin, *ibid.*, 391, 393 (the reign of Ashurnasirpal II), 406–407 (the reign of Tiglath-pileser III). On breaching the body of the wall, see *ANEP* 369 (the reign of Tiglath-pileser III).

II in Calah, which depicts the defenders attempting to seize the top of the battering ram close to the wall and to pull it with chains, while two of the attacking soldiers are hanging on the pole (or on the chain) as a counterweight to those trying to pull the ram upwards.[171]

(c) Penetration: the Entry of Warriors through the Breach

The counter-ramp that was discovered inside the wall of Lachish, opposite the Assyrian siege ramp, illustrates another aspect involved in combat against a fortified city: once the besieged became aware of the direction of the ramp under construction, and had estimated (based on the speed of construction and its progress) that it would reach the wall, they began building an inner counter-ramp. The purpose of the inner ramp seems to have been to serve as the basis for "going up/standing in the breach," as will be explained below.

The width of the siege ramp where it touches the wall of Lachish is about 25 meters; consequently, the assumed width of the battlefront, the section of the wall against which the battering rams were operated, was approximately 15 meters. After the attackers succeeded in breaching the entire width of this section with the rams, they had an opening through which they could break into the city. At that stage, all the defenders could do was try to stop the enemy from the counter-ramp, which had been prepared in advance along the breached section of the wall. The time required for constructing such a ramp was generally longer than that required for breaching the wall. Therefore, the inner ramp must have already been under construction before the battering rams were positioned, that is, before the external ramp reached the wall. The access difficulties in the built-up area being attacked, [172] combined with time pressures, often made it impossible

171 See Yadin, *ibid.*, 393. For neutralization of siege engines used against various parts of the wall by seizing them with looped ropes and tilting them, compare Thucydides 2.76; Aeneas the Tactician 32.4.

172 Compare the instruction given by Philo of Byzantium to leave a space of about 60 cubits (26.6 meters) between the wall and the houses of the city, to facilitate the carrying of stones and reinforcements to any section of the wall (*Poliorcetica*, eds. Diels and Schramm, 10). The crowded building in most cities of the ancient Near East did not match this instruction.

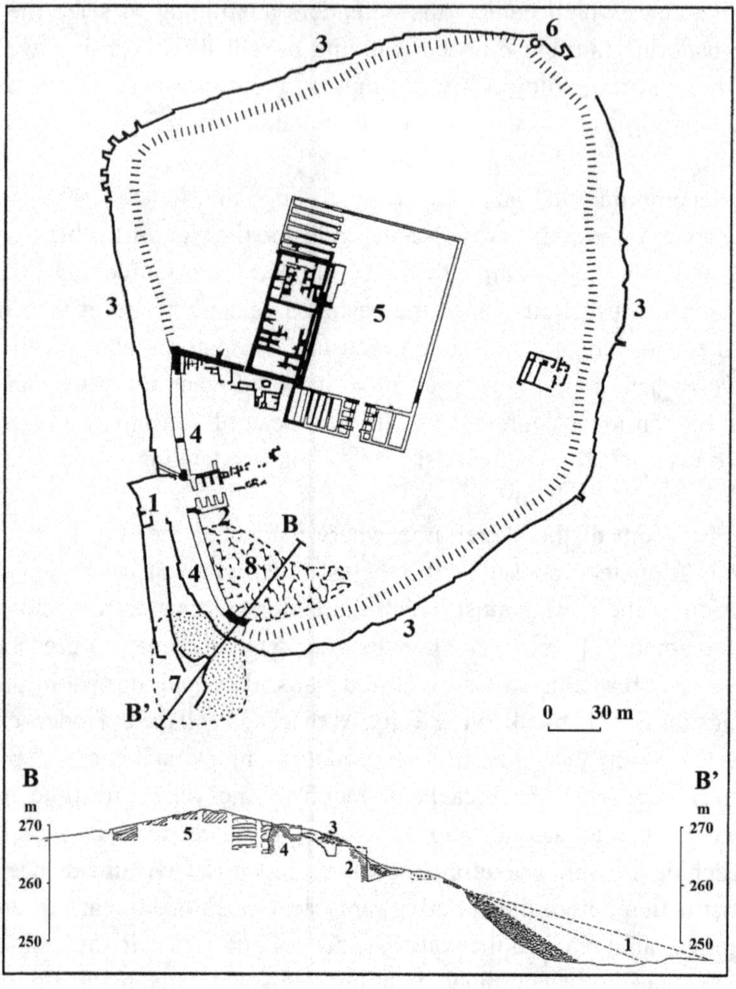

Fig. 9: Above: Map of Lachish during Sennacherib's campaign against it. 1. Outer gate; 2. Inner gate; 3. Outer revetment wall; 4. Main city wall; 5. Palace-fort complex; 6. Well; 7. Siege ramp; 8. Counter-ramp. Below: Cross-section of the siege ramp: 1. siege ramp (two possibilities); 2. Tower on the outer revetment wall; 3. Siege ramp, second stage; 4. Main wall; 5. Counter-ramp Based on Ussishkin, *EI* 20 (1989), 97–114

for the defenders to be selective about the materials used in constructing the inner ramp. They were forced to use whatever came to hand, hoping to minimize the carrying distance and to complete the ramp expeditiously, using as few porters as possible. The circumstances and purpose of laying down a counter-ramp are reflected in the words of Jeremiah: "For thus said the Lord, the God of Israel, concerning the houses of the city and the palaces of the kings of Judah that were torn down against the siege ramps and against the sword, and were filled with those who went to fight the Chaldeans – with the corpses of men whom I struck down in my anger and rage, hiding my face from this city because of their wickedness" (Jer. 33:4-5).[173]

Many scholars also relate this stage of combat to the statement in Isa. 22:9-11: "And you took note of the many breaches in the City of David, and you collected the water of the Lower Pool; and you counted the houses of Jerusalem, and pulled down houses to fortify the wall; and you made a reservoir between the two walls for the water of the Old Pool." However, the waterworks mentioned suggest that this passage does not refer to measures taken during a battle over Jerusalem, but rather to preparatory activities, naturally extended over a longer period, that were carried out before Sennacherib's army even arrived in Jerusalem (for a similar description, see 2 Chron. 32:2-5, which is only an interpretation of accounts in the books of Kings and Isaiah rather than a claim to being an actual fact).

The Hebrew term for a breach in a wall is פרץ (Isa. 30:13; Neh. 6:1; cf. 2 Kgs. 14:13; Prov. 25:28; Neh. 1:3; 3:35; 4:1; 2 Chron. 26:6; 32:5). Confronting invaders who attempt to penetrate via the breach in the city wall is characterized by the Hebrew idioms known as עמד בפרץ ("standing in the breach") and עלה בפרץ ("going up/climbing into the breach"). These idioms have a clear military context, illustrated by the counter-ramp found at Lachish, echoed in Ezekiel's admonition to the prophets of Israel, those "degenerate prophets who prophesy out of their own imagination": "Your

173 Compare Thucydides 2.75 about the wall built by the people of Plataea over the city walls, in order to raise it above the siege rampart of the Lacedaemonians that was coming closer to the city walls. The additional wall was built quickly out of bricks taken from adjacent houses.

prophets, O Israel, have been like jackals among ruins! You have not gone up into the breaches and made a barricade (גדר, lit. 'fence') about the house of Israel so that they might stand in battle on the day of the Lord" (Ezek. 13:4, 5). In similar fashion, the prophet complains about the leaders of the people (the prophets, the priests, the princes and the "People of the Land") on whose account God has punished the land: "I sought a man among them who could build a barricade (גֹּדֵר גָּדֵר), who could stand in the breach (עֹמֵד בפרץ) before me to defend the land from ruin, but I found none" (Ezek. 22:30). A more distant metaphor is reflected in the phrase, "He would have destroyed them had not Moses His chosen one confronted Him in the breach (עמד בפרץ) to avert His destructive wrath" (Ps. 106:23). The nature of this stand in the breach is further clarified by the expression גָּדֹר גָּדֵר ("making a fence/building up a barricade"; Ezek. 13:5; 22:30) which, in the present context, refers to blocking the breach by setting up a barricade,[174] as well as by the use of the verb שחת ("to ruin, destroy") in Ezek. 22:30; Ps. 106:23 (on the use of this verb in the context of destroying fortifications, cf. Jer. 48:18; Ezek. 26:4; Lam. 2:5–8; 2 Sam. 20:15).

Once the wall was breached, both sides made haste, one to invade the city[175] and the other to block the gap. This stage involved face to face combat with neither side having any real cover. The result of this combat depended to a large extent on the courage and expertise of the combatants. At this difficult time, the leaders of the defenders were called upon to go into the breach and sweep others along, by personal example, in a final effort against the enemy.

At first glance, the effectiveness of building a counter-rampart at great effort seems puzzling, as it would only be of use after the decisive stage of breaching the wall. However, this difficulty is resolved if one recalls that in imperial wars, time was a decisive factor in the considerations of both

174 On this phenomenon compare, e.g., *Jewish Wars* 3.7.23–29 (the battle of Jotapata). On the idiom גדר פרץ in the sense of rebuilding walls which had collapsed, not necessarily in a military context, cf. Amos 9:11; Isa. 58:12; and similarly פרץ גדר, also not in a military context, in Isa. 5:5; Eccl. 10:8.

175 Compare, e.g., *Jewish Wars* 4.1.7 (the battle of Gamala).

parties,[176] therefore, once the attackers were convinced of the defenders' determination to continue the struggle, an opportunity was opened for negotiations, in which the defenders could expect better conditions than those that would await them should the city fall to the enemy.[177]

4. Siege Towers

The lay of the land did not usually provide the attackers with advantages of height that would facilitate observation and shooting at the wall and into the besieged city. In order to obtain the advantage of height, they resorted to artificial means. The use of siege towers eliminated the need for extensive earthworks for this purpose;[178] in certain cases, the towers were moved from one sector to another, providing the attackers some flexibility and facilitating the shifting of *Schwerpunkt* (pressure point) in the battle over the besieged city. These were moveable[179] wooden towers built in varying sizes.[180] The towers were probably assembled beyond the range of the defenders' fire and only brought close to the wall later.[181]

Siege towers are not shown in any reliefs from the ancient Near East. Akkadian sources do not explicitly state what was done with the towers

176 See Ephʿal in *History, Historiography and Interpretation,* 96–104; and also below, pp. 106–113.

177 On negotiations conducted after breaching the wall but before the final conquest of the city, compare Josephus, *Jewish Wars* 5.9.2–3; 6.6.2 (the Battle of Jerusalem).

178 An example of such earthworks is the artificial mound near the wall of Old Smyrna, that rises to a height of 11 meters above the city, which was destroyed close to the end of the 7th century BC. The excavators of the site believe that the mound served as a base for shooting arrows and stones into the city during a siege. They attribute the construction of this mound to King Alyattes of Lydia, who waged battle against Smyrna and conquered it. See Cook, *BSA* 53–54 (1958–59), 23–27; Nicholls, *ibid.*, 88–91, 128–134.

179 See: ARM II 7:8ff.; Shemshara 7:15–17.

180 Compare Grayson, *Chron.* 5:22: *[...]x iddi ṣāpāti rabûti*[meš] *ušbalk[it],* "(Nebuchadnezzar) encamped… large siege towers he moved acr[oss]."

181 Compare Grayson, *Chron.* 3:35–36: *[ina muḫḫ]i* [uru]*Anati ittadi ṣāpīt[u... ultu(?) e] berti ereb šamši [...]* KIR *ṣāpītu ana dūri uqtarrib ṣaltu ana... libbi [āli] ipušma,* "[Nabopolassar] laid siege to the city of Anat. A tower from(?) the western bank… he brought the tower close to the wall. He conducted an assault upon [the city and captured it]."

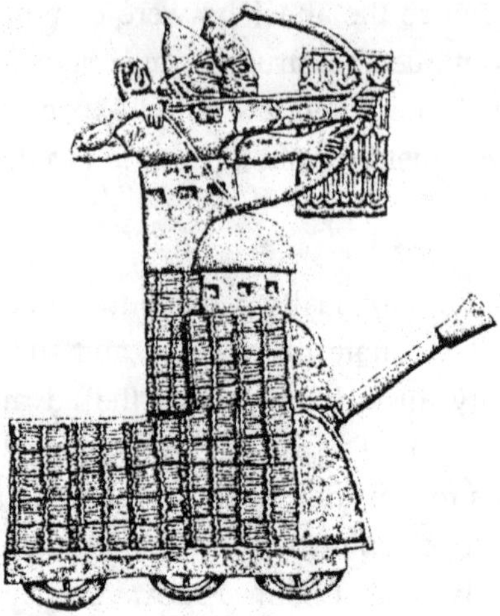

Fig. 10: A siege tower alongside a battering ram.
From the palace of Ashurnasirpal II at Calah (Nimrud)
From Yadin, "Weapons, Weaponry," *Enc. Miqr* 5.971–972

after they were brought in the vicinity of the city walls.[182] According to the Pi(ᶜankh)y Inscription, l. 32, archers and slingers were stationed on a wooden tower during the siege of Hermopolis, and from there shot and killed people in the city.[183] From the beginning of the 4th century BC onward, siege towers were also equipped with artillery devices.[184] Greek sources enumerate methods used against siege towers: a sortie from the city to set fire to the towers[185] (a method that was presumably effective in

182 A wooden device on wheels similar to a tower, with an Egyptian soldier standing in it holding a shield and an axe, is shown in a painting on a wall of a grave in the Asasif area in Thebes dated to the XI Dynasty, which depicts an attack of Egyptian soldiers on an Asian fortress or city. See: Jaroš-Deckert, *Das Grab des Jnj-jtj.f: Die Wandmalereien der XI. Dynastie*, 37–44. This wall painting contains no indication that the tower was used as a platform for shooting into the city or for storming it.

183 See: Lichtheim, *Ancient Egyptian Literature,* Vol. III*: The Late Kingdom,* 72.

184 Arrian, 1.20.8. On towers as platforms for the defenders' artillery, see 2 Chron. 26:14–15; see also belows, pp. 100–102.

185 Arrian, 1.20.9–10.

earlier periods as well) and stretching screens above the walls in order to stop projectiles from flying into the city.[186]

There are several known terms used to designate siege towers:

1. ^{giš}*dimtu*: found in sources from the first half of the second millennium BC (the Mari and Shemshara letters, the story of the siege of Urshu and the Babylonian *tamītu* texts);[187]
2. $^{(giš)}$*ṣāpītu:* found in the inscriptions of Ashurnasirpal II and in the Babylonian chronicles about the period of Nabopolassar and Nebuchadnezzar.[188]
3. *b3k* with the determinative "wood," found in the Pi(ʿanhk)y Inscription, ll. 32, 91.[189]
4. πύργος: found in Greek sources from the beginning of the 4th century BC onward.

The terms *dimtu* (OB onward) and πύργος were originally used to refer to stone or brick towers in the wall or to free-standing structures in open territory; their use in reference to portable siege equipment is secondary.[190] In order to distinguish between the two, Akkadian scribes added the determinative GIŠ, "wood" to the word for "siege tower" to indicate both the tower's use and the material from which it was built. The term $^{(giš)}$ *ṣāpītu* is borrowed from the Aramaic צפה ("to look, look out," signifying the purpose of the tower: observation) and, as noted above, is known from the first half of the 9th century BC. From then on, the use of the term *dimtu* continued in its original sense, while the borrowed term *ṣāpītu* was used specifically for the military device.

186 Diodorus, 17.25.5 and 17.45.3–4.

187 SH. 915:15 (*Shemshara* No. 7); Güterbock, *ZA* 44 (1938), 118:29', 32'; Beckman, *JCS* 47 (1995), 24; ARM XV, 198 sub *dimtum*; IM 67692:53.

188 RIMA 2 216:53; Grayson, *Chron*. 3:35–36; *Chron*. 5:22, respectively.

189 On the nature of this device, and discussion of the views of its significance, see Darnell, in *Ägypten im afro-orientalischen Kontext*, 73–76.

190 Cf. CAD D 144–146 *dimtu,* 1; *AHw* 170–171 *dimtu, 1*; Liddell and Scott, *A Greek-English Lexicon,* 1556 πύργος, 1. On the relationship between *dimtu* and πύργος in the non-military sense, see Heltzer, *JNSL* 7 (1979), 31–35.

5. Siege and Military Engines: nēpešu, μεχαναί, "clever devices"

Among the devices used in the conquest of a city, the inscriptions of Tiglath-pileser I, Ashurnasirpal II, and Tiglath-pileser III mention the term *nēpešū,* which was probably used to refer to battle equipment or machinery.[191] The Synchronistic History relates that King Nebuchadnezzar I of Babylon was forced to burn his *nēpešū* (which he brought with him when he attacked the Assyrian fortress of Zanuqu), in order to prevent them from falling into the hands of his enemy, King Ashur-resh-ishi I of Assyria, who was progressing toward him with his chariots.[192] It follows that these devices were flammable and slow to transport, so they could not be saved in the event of retreat. The term *nēpešū* is derived from the verb *epēšu* ("to act, to construct, to manufacture"), meaning "an [engineering] device/machine."[193]

In Greek and Hellenistic historiography from the latter half of the 5th century BC onward, the term μεχαναί is used as a general term for artillery devices and other engines of war that were used by attackers and defenders of cities alike.[194] Among these, particular importance was given to mobile siege towers, on which were stationed archers and various devices for hurling large stones and arrows and from which battering rams were operating to breach the walls.[195]

2 Chron. 26:15 states that King Uzziah of Judah made in Jerusalem "clever devices (חשבנות מחשבת חושב; lit., "contrivances, the invention of inventive men"), to be on the towers and the corners, to shoot arrows and great stones." The Septuagint translates חשבונות מחשבת חושב by the term μεχαναί, indicating that this Hebrew *hapax legomenon* was understood as an artillery device. Yadin rejected the opinion that the term חשבונות refers

191 Tiglath-pileser I: RIMA 2 34:34 (in the land of Qumanu); Ashurnasirpal II: RIMA 2 220:111 (against Udu); Tiglath-pileser III: *ITP* 162:21 (against Dur Balihaya, the capital of Bit Shaʾalli).

192 Grayson, *Chron.* 21 ii 2'-6'.

193 CAD N/II 170 3,b *nēpešu:* "tool, utensil, implement."

194 Liddel and Scott, *A Greek-English Lexicon,* 1131, μεχανή: "contrivance, machine, engine of war."

195 See Bar-Kochva, *Judas Maccabaeus: The Jewish Struggle against the Seleucids,* 21, 308, 317-323, and the detailed references therein.

to a device shooting arrows and stones, and argued instead that artillery was invented no earlier than the 4th century BC and therefore the attribution of sophisticated weapons that hurl large stones to the period of Uzziah (the 8th century BC) is clearly anachronistic.[196] On the assumption that the description in the Book of Chronicles is based on authentic historical knowledge of the period of Uzziah, Yadin suggested identifying the term חשבונות with special structures, protected balconies supported by beams constructed in crisscross fashion above the towers on the city walls, which projected outwards and were wider than the towers themselves, allowing slingers and archers to shoot without being excessively exposed (such structures appear in artistic descriptions from the 8th century BC).[197]

Yadin's comments on the invention of artillery seem plausible,[198] but my conclusion is different. The semantic similarity between the terms μεχαναί and חשבונות מחשבת חושב, suggests that the Chronicler invented this term to refer, anachronistically, to the artillery devices that were common in his own day (ca. 400 BC), and therefore (clumsily) translated a parallel term that was absent in Hebrew.[199] Moreover, according to 2 Chron. 26:15, the "clever devices" that were built upon the towers and on the corners, were intended "to shoot arrows and *great stones*" (as opposed to the small "stones for slinging" mentioned in verse 14, that could be thrown

196 Sukenik (Yadin), *BJPES* 13 (1946-1947), 19-24; Yadin, *Art of Warfare,* 326-327.

197 Clear representations of the balconies under discussion can be seen in the relief depicting the battle of Lachish. See Ussishkin, *The Conquest of Lachish by Sennacherib,* segments III-IV (see also the enlarged photographs there, 103, 106). For a bronze model of such a structure, found in Urartu (BM 91250), see Boardman, *Cambridge Ancient History, Plates to Volume III*, 60, fig. 92.

198 See below, p. 104.

199 Welten's awareness of the semantic resemblance between the terms μεχαναί and חשבונות מחשבת חושב led him to the conclusion that the Book of Chronicles was written in the last third of the 4th century BC, at the beginning of the Hellenistic period (see Welten, *Geschichte und Geschichtsdarstellung in den Chronikbüchern,* 111-114). However, this is not necessarily correct: the use of artillery was already known in the eastern Mediterranean basin from the beginning of the 4th century BC, and possibly one or two generations earlier (see p. 104). It is therefore plausible that the need to invent a term for artillery weapons existed from the time of their appearance in the area, even before its conquest by Alexander the Great. See further below, p. 105, n. 206.

by hand). The height of the balconies or projections referred to by Yadin was insufficient to significantly augment the range of the larger stones if they were thrown by hand rather than by a special device. Consequently, Yadin's interpretation of the phrase does not seem reasonable.

In conclusion, the vague and undefined derivation of the terms used in the three languages suggests that they do not refer to a specific, well-defined weapon, but rather serve as a general term for siege and war engines. The specific meaning changed in accordance with the technique of warfare and its circumstances.

6. Strategems

Every series of possibilities detailed in the Assyrian queries of prevailing over a city, whether by negotiations or by force, concludes with the question "[Will the city be caught] by any strategem?" – that is, in an unexpected manner which would not occur to anyone and therefore could not be specified explicitly in the query. Many conquest traditions refer to various strategems, some which may have been practical and others that seem completely imaginary.

The archetypal imaginative strategem was the infiltration of Greek warriors into Troy in the wooden horse, after they had failed to overcome the city over a long period of time (*Odyssey* 8.487–520).[200] The same motif appears in the Egyptian story of the capture of Jaffa in the period of Thutmosis III by sneaking soldiers into it, hidden inside baskets.[201]

200 The story of the Iliad takes places in the tenth year of the battle of the Achaeans against Troy (see *Iliad* 2.134, 295). During the story, which continues for 52 days, four battles took place near Troy.

201 See Wilson, *ANET,* 222–223. On the literary nature of this story, and parallels of the hero's tactics with several of the adventures of Odysseus, see Goedicke, *Chronique d'Égypte* 43 (1968), 219–233. A letter sent to King Zimri-Lim of Mari (A.319: 36ff.) tells of a failed strategem: Atamrum, king of Allahad, who had laid siege to the city of Razama and found it difficult to overcome, armed thirty "criminals" with lances (it is unclear what his intention was by this: perhaps to make the besieged think that these were their supporters, neighbors or soldiers of Zimri-Lim coming to their aid, and therefore to open the city gates to them). The text is not sufficiently clear, but the context suggests some sort of trick. On the document in question, see Charpin, *MARI* 7 (1993), 197–203.

Sophisticated strategems are also attributed to historical figures: Pausanias relates that Solon (early 6th century BC) placed hellebore roots in the water supply of the city of Cirrha, the port city of Delphi, and infected its inhabitants and defenders with severe diarrhea, forcing them to abandon their posts on the walls of the besieged city.[202] Herodotus tells of the capture of the city of Barca by Amasis (the general of Darius I) after nine months of exhausting siege by making a treaty with its inhabitants that was to remain valid "so long as this land shall stand," when in fact the treaty was made above a pit that had been dug by the Persians and concealed by planks covered with earth (4.201).

One of the strategems used in the traditions of siege entailed tricking the defenders to go outside of the city by staging a retreat or withdrawal of the attackers. The defenders, thinking that their enemies were withdrawing due to weakness, move out of the city and chase the retreating forces. At this stage, a second group of attackers, "the ambush," that remained hidden until that point, enters and attacks the city. The Bible mentions two cases, quite similar in formulation, in which this tactic was used: the conquest of the town of Ai by Joshua (Josh. 8:2–22, 24) and the conquest of the town of Gibeah during the war of the Israelite tribes against Benjamin (Jud. 20:29–42, 48). Parallels are also found in ancient military literature.[203] Suspicion of such a strategem is implied in the reaction of the Israelite king upon hearing that the Aramean camp, that had laid siege to Samaria, was empty of soldiers, "And the king rose in the night and said to the servants, 'I will tell you what the Arameans have prepared against us. They know that we are hungry: therefore they have gone out of the camp to hide themselves in the open country, thinking, "When they come out of the city we shall take them alive and get into the city"'" (2 Kgs. 7:12).

7. The Impact of Artillery on Siege Warfare

A decisive change in the technique of siege warfare occurred with the invention of artillery, a class of weaponry which enabled the shooting of

202 Pausanias, *Description of Greece,* 10.37.7.

203 Frontinus, *Strategems,* 3.10.1–9. For an extended discussion of this strategem, see Malamat, *History of Biblical Israel,* 86–91.

rocks or other projectiles with greater force than possible by a single person. This system was developed first and foremost for siege warfare, and was not used in field battles for a long period. The emergence of artillery was marked by the development of the *gastraphetes*, a shooting device that worked like a large, composite bow that was pulled back mechanically and shot spears. Within a few decades, once the technique of torsion was introduced, such shooting devices (catapults) became far more powerful, making it possible to shoot spears and stones of a far greater weight and over distances significantly greater than those known until then.[204]

According to the written sources, these war engines were initially developed in Syracuse (in the rule of Dionysius I, who laid siege to Motya in 397 BC) and in Macedonia (in the reign of Philip II, who attacked Perinthus in 340 BC, and his son Alexander, who battled against Halicarnassus in 334 BC and against Tyre and Gaza in 332 BC).[205] In the sieges of Motya, Perinthus and Halicarnassus, artillery was only operated against people: by the attackers, in order to neutralize the defenders on the walls and stop forays from the city against the besiegers, and by the defenders, in order to keep the attackers away from the walls. Such a use of artillery was purely technical and did not involve any fundamental innovation in military thinking, as it maintained the routine goals of creating fire (albeit more effectively), while the task of breakthrough remained the domain of battering rams and tunnels, as in the past. The first fundamental innovation is found in Alexander the Great's siege of Tyre. Here, in addition to previous tasks, artillery was also used to penetrate the wall (in its weaker sections),

204 For a description of these shooting machines, including details of the sources discussing their performance and the stages of invention and development of the different models, see Marsden, *Greek and Roman Artillery: Historical Development,* 48–98.

205 In the excavations at Kouklia (Palaepaphos), large heavy stones were discovered in the rampart adjacent to the gate of the city. According to Maier, this rampart was laid down in the 5th century BC (see above, pp. 28–31). This chronological determination leads to the conclusion that the invention of artillery preceded by at least a 100 years what has been found in the written documentation. For a refutation of the theory that artillery devices were invented by the Persians and already used in the 6th–5th centuries BC, see Pimouguet-Pedarros, *REA* 102 (2000), 5–26.

so stone-hurling catapults carried out the task of the battering rams.[206] From this point onwards, a series of significant changes took place in siege warfare, the most important characteristics of which were the following:

1. The ability to damage walls and towers rapidly and from a distance greatly reduced although it did not completely eliminate the need for earthworks (siege ramps and tunnels) and battering rams, and in many cases made it possible to shorten the duration of the siege.
2. The depth of the field of fire, which had until that point been roughly equal for both attackers and defenders – as it was dependent on the range of bows and slings and on the individual human ability of the warriors – was from now on determined by mechanical means, which could vary considerably in their performance. The use of different models of artillery devices could significantly change the depth of the field of fire for either of the combating sides.
3. In general, the danger to the noncombatant population in the city increased radically. Extension of the target of projectiles increased the number of casualties and thereby affected the resistance power of a besieged city.

Like most military inventions, the use of artillery probably spread rapidly and was also adopted in the Near East. From reports concerning the use of firing devices by the defenders of Halicarnassus, Tyre and Gaza[207] against Alexander the Great, it would appear that the innovation of artillery was introduced into the Persian Empire (which controlled those cities) as well, once its advantages were discovered. It is unclear whether they succeeded in adopting the technique of torsion, which was invented around 350 BC, and to use it in their battles against Philip II and Alexander, who already had at their disposal sophisticated shooting devices utilizing this technique.[208]

206 On the use of artillery in the five sieges mentioned here, see Marsden, *ibid.*, 99-104. According to Diodorus, 17.43.4, in the battle of Tyre, Alexander's densely positioned artillery created a breach one *plethron* (=100 feet) wide in a wall whose thickness (together with the additions added to it prior to the battle) was no less than 30 feet.

207 Halicarnassus (Diodorus, 17.24.6; 26.6); Tyre (Arrian .2.21.3; compare Diodorus 17.41.3-4; 42.1); Gaza (Arrian 2.27.2).

208 See Marsden, *ibid.* (above, n. 206), and the table, *ibid.*, p. 43.

8. Considerations for Preferring Siege to Surrender

The extant sources generally demonstrate the suffering of the besieged; however, it is important to recall that the situation of the besiegers was not much better in most cases. In addition to the efforts needed for the breakthrough itself, the besiegers had other difficulties to deal with. These ranged from the gathering of food (even if they lived off the land) and its transportation, which involved considerable efforts that increased as the siege continued and the lines of supply became longer, through the difficulties of obtaining water in areas where flowing water was not readily available (as the besieged inhabitants attempted to withhold water from their attackers or make its acquisition difficult),[209] to the difficulties of extended stay on the battlefield and the arduous living conditions in camps, exposed to the vagaries of weather.[210]

The besiegers needed to protect themselves against enemy attack from two different directions: from within the besieged city and from outside. Sorties outside the city walls were intended, first and foremost, to damage

209 See, e.g., the description of Hezekiah's preparations for war against Sennacherib: "And when Hezekiah saw that Sennacherib had come and intended to fight against Jerusalem, he planned with his officers and his mighty men to stop the water of the springs that were outside the city; and they helped him. A great many people were gathered, and they stopped all the springs and the brook that flowed through the land, saying, 'Why should the kings of Assyria come and find much water?'" (2 Chron. 32:2-4). The impact of redirecting the waters of the Gihon spring into the city must have been severe, and was probably particularly felt during the protracted Babylonian siege of Jerusalem in 588-586 BC. On fetching water to supply the besiegers of Razama from a great distance, an hour's walk, with the concomitant danger of harm to the water bearers, see A.319:47-48 (for a transliteration and translation of this letter and a discussion of its contents, see Charpin, *MARI* 7 [1993], 197-203).

210 Compare the words of Uriah the Hittite to David: "The ark and Israel and Judah dwell in booths (ישבים בַּסֻּכּוֹת); and my lord Joab and the servants of my lord are camping in the open field; shall I then go to my house to eat and to drink, and to lie with my wife?" (2 Sam. 11:11); "...Ben-hadad was drinking himself drunk in the booths (שֹׁתה שכור בַּסֻּכּוֹת), he and the thirty-two kings who helped him" (1 Kgs. 20:16; cf. *ibid.*, v. 12). I do not accept Yadin's theory (*Biblica* 36 [1965], 332-351) that סכות in these passages refers to the toponym of the town of Succoth in Transjordan. On the dwelling of the besiegers in tents during the war against Samaria and Jerusalem, compare 2 Kgs. 7:7-8; Jer. 37:10, respectively. On the vagaries of climate, see "The General's Letter" (below, p. 109, n. 221), ll. 21-25, 27-28.

siege installations, and at times were also aimed against enemy forces who were engaged in setting up or operating these installations. The purpose of the sortie was not to bring about an all-out confrontation in field battle between all combatants on both sides,[211] as the defenders were not interested *ab initio* in a confrontation of this type, but only in preventing the operation of the siege equipment and in thwarting the enemy's approach to the wall. In this respect, the sortie may be described as a simultaneous counter-attack.[212] The sortieing warriors set fire to siege towers and attacked the besiegers who came to extinguish the fires.[213] The losses caused to the forces of the surprised besiegers were likely to be so heavy that, at times, they were forced to withdraw and stop the siege.[214] At times, the sortieing

211 On a confrontation near the city after the orderly alignment of the forces of both sides (as opposed to a battle which developed after a sortie of the defenders during a siege), compare 2 Sam. 10:7–14 // 1 Chron. 19:8–15 (the battle of Abishai, brother of Joab ben Zeruiah, against the Ammonites); Diodorus, 17.11–12 (Alexander's war against Thebes). Mari Letter A.319 ll. 8–10 states that when Atamrum, King of Allahad and his army arrived at Razama, the people of the city came out and killed 1,300 of them (on this fascinating document, see above, n. 209). It is possible that this is not a reference to a sortie either, but rather to a field battle that preceded the siege.

212 Mari Letter A.319 (see above, n. 209) relates that after the failure of negotiations between the elders of Razama and Atamrum who laid siege to the city, the people of the city used to sortie and kill Eshnunna combatants in the besiegers' camp (ll. 23–24); and when the siege rampart reached the lower city, the besieged made a sortie early in the morning and killed half of the soldiers (of all the besieging force or perhaps only of the units building the rampart; ll. 29–31); cf. *ibid.*, ll. 47–51 (on sorties and attacks on the water bearers to the besiegers).

213 See, e.g., Diodorus, 17.24.5, 26.1–27.3 (Alexander's war against Halicarnassus; on the sorties during that battle, see also Arrian 1.20.4, 9–10, 21.5, 22.1–7); Arrian 2.27.1–2 (Alexander's war against Gaza). On the burning of the rampart laid by Esarhaddon against Uppume, the capital of Shubria, in the middle of the night, see Borger, *Asarh.* 104 ii 3–7 (the incident took place on the night of 21st Kislev, when the moon rises at 23.35 and is approximately one-quarter full); on the burning of the siege tower of the Elamites who were waging war against the city of Hiritum, see ARM XXVII 141. Cf. below, n. 214, on naval combat.

214 See, e.g., 2 Sam. 11:23–24 (the siege on Rabbath-ammon in the reign of David); 1 Kgs. 20:13–20 (the siege of Samaria in the reign of Ahab); YOS X 24:8: *nakrum ālam ilawīma ālum uṣṣâmma nakram isakkip* ("the enemy will lay siege upon the city, but the city shall raid and repulse the enemy").
For naval parallels to land sorties, see: Diodorus, 17.42.1–4; Arrian, 2.19.1–5,

warriors, if they were not local inhabitants, even planned to abandon the starving city and to get away from it.[215] Sorties were generally conducted at times when the besiegers were likely to be resting from their hard labor in building the siege facilities and from the activity involved in drawing close to the wall:[216] at night, at dawn or at noontime.[217] The walls and the buildings within gave the defenders a distinct advantage: they were able to observe the enemy and his activity, sometimes even at a great distance, and to benefit from a reasonable reaction time, while the besiegers could not know what was going on within the city.[218] The absence of adequate advance warning of the *time* of the sortie, its *magnitude* and at times even its *exact location*,[219] forced the besiegers to maintain high levels of alertness day and night, involving relatively large forces.

21.6–22.5; Curtius Rufus, 4.2.21–22, 3.10, 24, 4.6–9 (Alexander's battle against Tyre); Thucydides, 1.117 (the Athenian siege of Samos).

215 Streck, *Asb.* 68 viii 35–39 (Arabs from Babylon during Ashurbanipal's battle against Shamash-shum-ukin); Thucydides, 3.20–24 (combatants from Plataea during the siege by the Peloponnesian army, 428 BC. The purpose of this sortie, which was conducted on a dark and rainy winter's night, was not to damage the facilities and camp of the besieging army, but to break through and escape from the starving city). Compare: "When the king of Moab saw that the battle was going against him, he took with him seven hundred swordsmen to break through, opposite the king of Edom; but they could not" (2 Kgs. 3:26).

216 On the difficulties involved in constructing a siege ramp and building siege facilities, compare Ezek. 29:18: "Son of Man, King Nebuchadrezzar of Babylon has made his army expend vast labor on Tyre; every head is rubbed bald and every shoulder rubbed bare. But neither he nor his army have had any return for the labor he expended on Tyre"; Borger, *Asarh.*, 104 i 36–37: *xxx a[ra]mmu ina šipik eprī iṣē u abnē marṣiš pašqiš [ušakbis]* ("... [I stamped down] a ramp, by gathering earth, trees and rocks, with labor and toil").

217 On night sorties, compare Diodorus, 17.24.5 (=Arrian 1.20:9–10); Thucydides, 5.115 (Melos; 416 BC); "The General's Letter" (below, n. 221), rev. 6'-10'; cf. also Thucydides, above, n. 215. On dawn sorties, cf.: A.319:29–31 (above, n. 212); Diodorus, 17.26.3–4. On sorties at noon, see 1 Kgs. 20:15 ff. and Arrian 2.21.8–22.2.

218 One method of overcoming this limitation of the attackers was, of course, the use of siege towers wherever possible; see pp. 97–99.

219 The most convenient place for a sortie of a large force (compare 2 Kgs. 3:26: "seven hundred swordsmen") was, of course, the city gate, which was opened wide to facilitate the rapid exit of combatants. If the besieging force succeeded in repulsing the raiders, the latter found it difficult to retreat in an organized fashion back into the city (even

Another danger to be expected was the arrival of reinforcements to help the besieged, as noted above. On more than one occasion, the troops who came to the aid of a besieged city defeated the besiegers.[220] If it was impossible to stop such reinforcements beyond the range of combat over the city,[221] the besiegers were forced on more than one occasion to stop the siege and redeploy all of their forces to an open confrontation in field battle with the reinforcements.[222] The interruption of the siege, even for short periods, enabled the besieged to renew their strength in various ways, such

---

through the gate), and there was a danger that the enemy would break through the gate in their wake; cf. Arrian 1.22.3–7; Aeneas, Ch. 39; 2 Sam. 11:23–24.

220 See the Pi(ʿankh)y Inscription, ll. 17–21 (the battle near Heracleopolis); compare Josh. 10:5, 9–10 (Joshua's attack on the five Amorite kings who besieged Gibeon); and also the omen TCL VI 1 rev. 55: "An auxiliary force will go to the city to which you lay siege." On the possible arrival of an Assyrian expeditionary force, under the command of the chief eunuch, to help the besieged fortress of Ṣiṣṣirtu in the land of Harhar at the border of the land of Elipi, see Esarhaddon's query to Shamash, SAA IV 78. On the anticipation by the inhabitants of the besieged city of Razama for the arrival of Zimri-Lim and his army to come to their aid, and preparations for this arrival in the enemy camp, see A.319:45–47, 52–54 (on this document, see above, n. 209).

221 Compare the "General's Letter," especially rev. 10'ff. The "General's Letter" from Ugarit, contemporary of the el-Amarna letters (ca. mid-14th century BC), is written in Babylonian. It was sent to a king whose exact identity is not explicitly stated; from the contents and the political context, he would appear to have been a Hittite king or a local king who ruled in Syria under the aegis of the Hittite king. The sender of the letter is an officer who was stationed somewhere in southern Lebanon with his unit, including charioteers, to blockade various cities in that area and to prevent Egyptian reinforcements from getting there. On the language of this document and its historical background, see Izre'el and Singer, *The General's Letter from Ugarit* (for an earlier discussion of this document, including discussion of the geographical background and a suggestion regarding its historic background, see Rainey, *UF* 3 [1971], 131–149). It follows from this document that the forces commanded by the sender were performing this mission for at least five months at that stage; the letter reflects the importance of maintaining an ongoing blockade on the one hand, and on the other, the impact that the protracted stay in the field had on the guarding unit, on its equipment, on the physical condition of the men and on their morale.

222 Cf. Jer. 37:5ff. (there is no evidence that the Egyptian army in fact arrived in Palestine and that Nebuchadnezzar battled against it); 2 Kgs. 19:8ff. (the battle between the army of Sennacherib and the Egyptian army was carried out in Eltekeh, on the coastal plain).

as by evacuating the non-combatant population, obtaining new materiel and reinforcements, replenishing food and water supplies, repairing and strengthening buildings and installations that had been damaged during the siege, destroying ramps and other siege facilities which the temporarily-retreating army was unable to take with it.[223] Consequently, a renewed siege would have to start from the beginning.

Experience indicates that not every siege on a city ended with its conquest. Several of the cities that withstood siege and were not conquered merit particular mention: Emar, which in the days of Pilsu-Dagan son of Baal-kabar, withstood the king of the Hurrians and his army,[224] Hadrach, that withstood the army of Barhadad, king of Aram and his allies,[225] Damascus, which was not conquered by Shalmaneser III, [226] Tushpa, the capital of Urartu, which did not fall to Tiglath-pileser III,[227] Tyre, which withstood Shalmaneser V and Nebuchadnezzar[228] and Jerusalem, that withstood Sennacherib's forces.[229] The dates found in Siege Documents 2NT 296

223 Cf. "The General's Letter" (above, n. 221), obv. 22–25, 29–30; Thucydides, 1.117; 5.115 (this was a night-sortie, which made it possible to bring grain and other necessities into the city); 2 Kgs. 7; Jer. 37:5ff.

224 Cf. *RPAE* 42:8–16; *TSABR* 9:21–22; see Tsukimoto, *ASJ* 12 (1990), 189–192, No. 7:29–33; and below, pp. 136–137.

225 See the inscription of King Zakkur of Hamath and Luʿash (*KAI* 202), from ca. 800 BC.

226 See RIMA 3 48:15'–16'. The inscriptions of Shalmaneser III relate that he blockaded Hazael, king of Aram, in Damascus and that he cut down his orchards, but there is no mention of the conquest of the city.

227 See *ITP* 124:23–24; 134:21'–25'. These inscriptions state that Tiglath-pileser killed many of the warriors of King Sardur of Urartu opposite the gates of Tushpa and even carved his royal image in the rock near the besieged city. However, there is no mention in Tiglath-pileser's inscriptions of the conquest of the city.

228 On Tyre's endurance against Shalmaneser V for five years, see Josephus, *Antiquities* 9.14.2; on Tyre's endurance in the time of Nebuchadnezzar, *ibid.*, 10.11.1; and *Against Apion,* 1.21 (according to the sources cited by Josephus, Nebuchadnezzar's siege of Tyre lasted 13 years); compare Ezek. 29:18.

229 Isa. 36–37 // 2 Kgs. 18:13, 17 – 19, 37 do not explicitly state that Jerusalem was subjected to a real siege. However, Isaiah's words ("Therefore thus says the Lord concerning the King of Assyria: He shall not come into this city nor shoot one arrow there, nor come before it with a shield nor cast up a siege ramp against it"; 2 Kgs. 19:32 // Isa. 37:33), suggest that there is no indication in the prophetic account

and 297 from Nippur indicate that the city withstood Nabopolassar for a period of at least four months;[230] the city of Assur withstood the siege of Nabopolassar from Iyyar 615 BC until at least day X of Sivan.[231] Other cities were captured only after extended sieges: Babylon, after it was besieged by Sennacherib for more than 15 months[232] and by Ashurbanipal for at least 22 months,[233] Samaria by Shalmaneser V only "at the end of three years,"[234]

---

that Sennacherib laid siege to Jerusalem. Likewise, the Sennacherib Inscription – "I imprisoned him (Hezekiah) in his royal city, Jerusalem, like a bird in a cage. I blocked him with barricades and made it unthinkable (lit. 'taboo') for him to exit by his city gate" (OIP II. 33 iii 27–30) – does not indicate any siege activity in order to break through (compare the description of Sennacherib's activities against the 46 fortified cities of Judah; *ibid*., 32–33 iii 18–23), but at most a blockade which was not longstanding – as follows from the fact that the entire "Third Campaign" lasted no more than a year. Hence, there is no need to exaggerate the seriousness of Sennacherib's attack on Jerusalem. On the historiographic and historic significance of the literary images by which the war against Judah is described in the Sennacherib inscriptions, see Tadmor, *Zion* 50 (1985), 65–80.

230 Oppenheim, *Iraq* 17 (1955) 87–88 (=O.23; 30 in the catalog of Brinkman and Kennedy; see below, p. 116, n. 5).

231 Grayson, *Chron.* 3:16–18.

232 See Brinkman, *JCS* 25 (1973), 93–94.

233 Ashurbanipal's siege of Babylon began on 11 Tammuz, year 18 of Shamash-shum-ukin (Grayson, *Chron.* 15:19). The latest date of the extant "siege documents" from that same siege is 29 Iyyar in the 20th year of his reign (San Nicolò, BR 8/7 No. 20 = K.140 in the Brinkman-Kennedy catalog).

234 2 Kgs. 17:5–6; 18:9–10. The Bible mentions only the name of Shalmaneser V as the Assyrian king who waged war against Samaria and besieged it. However, from what is recorded in the Babylonian Chronicle concerning the accomplishments of Shalmaneser V and from the Sargon inscriptions, it follows that Samaria was conquered *twice*, once by Shalmaneser V and once by Sargon. Reconstruction of the events on the basis of all of these data entails issues of interpretation and evaluation of the historical sources under discussion. On the view that Samaria was conquered twice: once in 722 BC, following a siege by Shalmaneser V which lasted no less than two calendar years (alluded to in the verse "and for three years he besieged it", see 2 Kgs. 17:5; compare 2 Kgs. 18:9–10: "Shalmaneser, King of Assyria came up against Samaria and besieged it and the end of three years he took it"), and a second time (in 720 BC) in the second year of Sargon's reign, see Tadmor, *JCS* 12 (1958), 33–40; for a summary of the research in light of this approach, see also Cogan and Tadmor, *II Kings,* 197–201. Regarding the view that Samaria was in fact conquered twice, by both of the Assyrian kings mentioned, but that it was not subject to a protracted siege, and that the verse referring to three years of siege is no more than a mistaken

Jerusalem by Nebuchadnezzar after 18 or 30 months,[235] Barca by Amasis (commander of the Persian force that fought in Cyrene during the reign of Darius I) at the end of nine months,[236] Samos by the Athenians under the command of Pericles in the ninth month of siege,[237] Megiddo by Thutmosis III at the end of seven months,[238] Tyre by Alexander in the seventh month of siege,[239] Sanahuitta by Hattusili I at the end of five months[240] and Soloi by the Persians in the fifth month of siege.[241] Letters from the late 19th and early 18th century BC mention three years of siege on uruŠušarrae(?) and nine years of siege on the city Harsamna.[242]

---

reconstruction by the Deuteronomistic historiographer based on the sources available to him, see Na'aman, *Biblica* 71 (1990), 206–225.

235 Cf. Jer. 39:1–2; 52:4–7; 2 Kgs. 25:1–4. On the calculation of time on the basis of the data in these writings, based on the various viewpoints concerning the beginning of the year according to the counting of the Judean kings, see Malamat, *IEJ* 18 (1968), 150 ff.

236 Herodotus, 4.200.

237 Thucydides, 1.117.

238 Gebel Barkal Stele, ll. 21–22 (*ANET,* 238), first half of 15th century BC.

239 Diodorus17.46.5; Curtius Rufus, 4.4.19; Plutarch, *Alexander*, 24.3.

240 4th year of Hattusili I; see Houwink ten Cate, *Anatolica* 11 (1984), 49:23 (see also *ibid.*, 67), mid-17th century BC.

241 Herodotus, 5.115 (498 BC).

242 SH 812:58–59: "Now, his master is under siege for three years (but) he has not come." Shemshara, No. 63; Balkan, *Letter of King Anum-hirbi of Mama to King Warshama of Kanish,* 7:29–31: "When your father Inar (king of Kanish) laid siege to Harsamna for nine years, did my land attack your land?" The results of the two sieges mentioned here are not known.

In most of the translations of the inscription from the tomb of Ahmose son of Ebana in el-Kab, l. 15, it is stated that during the war of Amosis I against the Hyksos, the city of Sharuhen was besieged for three years. For the view that this inscription refers to activities that were carried out in the course of three years, and not continuously throughout this period – a view based on both linguistic and logistic-military considerations – see Goedicke, *JARCE* 11 (1974), 40–41; *idem.*, in *Egyptological Studies in Honor of R. A. Parker,* 42–44. According to Herodotus 2.157, Ashdod was conquered by King Psammetichus (I) of Egypt following a siege that lasted 29 years: "Among all the cities, the one that exhibited the greatest forbearance under siege, according to what is known to us, for the longest period of time, was this Ashdod." Most scholars hold that Herodotus' remarks concerning the extraordinary length of that siege make no sense. Accordingly, it has been suggested that the statement about the 29-year siege of Ashdod is associated with other data in Herodotus' book, i.e.,

Consequently, the rationale motivating defenders to prefer siege above surrender, from the onset or during the course of battle, was the evaluation of the difficulties that confronted the besiegers, in light of the limitations of time and space that restricted the operation of imperial armies against cities of the ancient Near East,[243] and in light of the knowledge that not every siege ended in conquest.

---

concerning Psammetichus' meeting with the Scythians, who passed through Ashkelon on their way toward Egypt (1.105), and concerning the Scythian rule in Asia, that lasted for 28 years, until they were defeated by Cyaxares, king of Media (1.106). For a summary of the research along these lines, see Na'aman, *TA* 18 (1991), 39-40 (and references therein). For another assumption, namely that it may be understood from Herodotus that Psammetichus' siege on Ashdod began in the 29th year of his reign, see Tadmor, *BA* 29 (1966), 101-102.

243 On these factors and their operative consequences, see Ephᶜal, in *History, Historiography and Interpretation*, 91-104.

CHAPTER FOUR

LEGAL AND ECONOMIC ASPECTS

I. INTRODUCTION

The food shortage and the impossibility of obtaining supplies from the outside created a unique economic situation within besieged cities, characterized by changes in the system of prices: a sharp increase in the price of food and a decrease in the cost of all other goods. Particular decrease was evident in the value of human-beings, primarily slaves (as their economic benefit diminished while agricultural work and other duties performed outside city walls were at a standstill, as were many of their duties within the walls). The value of children decreased as well. The longer the siege continued, the blessing generally associated with numerous offspring was turned into a curse, and people were forced to sell their children (or even give them away for free) in order to "keep them alive" (*bulluṭu*).

The value of real-estate probably suffered as well in times of siege. However, unlike slaves or grain, whose absolute prices are known, we have no information on which to evaluate fluctuations in real-estate prices. While it is reasonable to assume that the value of property continued to decrease the longer the siege continued, there were always people who were willing to purchase property at bargain prices in the hope of realizing a large profit on their investment after the siege.[1] These buyers made sure that their property transactions were documented according to proper legal

1 The record of a house sale "in a year of hardship", *TSABR* 65 from Emar, provides explicit evidence of the buyers' expectation of a 100% profit on real-estate at the end of the hardship period (see below, p. 146, n. 88).

procedure (cf., Jer. 32:9–14;[2] and the discussion of the legal significance of the hardship formulae, below, pp. 143–147).[3] Jeremiah's comments (32:24–25) seem to imply that the prophet had some doubts as to the economic sense of buying the field in Anathoth from his cousin, Hanamel, during an advanced stage of the siege on Jerusalem, when the city was facing certain destruction and exile of its inhabitants.[4] Nevertheless, when ordered to place the documents ("the deed of purchase," the "sealed deed," and the "open deed") "in an earthenware vessel, so that they may last for a long time" (*ibid.*, v. 14), his act was symbolic of the hope for the distant future: "For thus says the Lord of Hosts, the God of Israel, houses and fields and vineyards shall again be bought in this land" (*ibid.*, v. 15; cf. vv. 42–44).

There are several dozen Babylonian documents extant from the 7th century BC (extending over a period of some 70 years, beginning from 690 BC) from the cities of Babylon, Nippur, Uruk and Dilbat(?), which

2 On the economic reality of siege, characterized by rapid increase in food prices and the sale of property (specifically real-estate) at bargain prices, cf., Ezek. 7:12–19.

3 Babylonian siege documents detailing real-estate are: H.2; K.119, 140 (references for these documents appear below, n. 5). For Emar documents detailing the sale of property in "the year of hardship/famine (and hostilities)": see *RPAE* 20, 111, 138, 139, 149, 158, 162; *TSABR* 57, 65; *ASJ* 12 (1990), 207–208, no. 15; 208–209, no. 16; *RE* 31.

4 Jer. 32:24–25: "Behold, the siege mounds have come up against the city to take it, and because of sword and famine and pestilence the city is given into the hands of the Chaldeans who are fighting against it. What you have spoken has come to pass, and behold, you see it. Yet You, O Lord God, have said to me, 'Buy the field for money and get witnesses' – though the city is given into the hands of the Chaldeans." On the decisive significance of this stage in the battle for Jerusalem, see below, pp. 84–90.

5 These documents are referred to below according to their numbering (beginning with the letters H, J, K, N, O, P and T) in Brinkman and Kennedy's catalogs of Babylonian economic documents from the years 721–626 BC, and in Kennedy's for the years 626–605; see Brinkman and Kennedy, *JCS* 35 (1983), 1–90; 38 (1986), 99–106; Kennedy *JCS* 38 (1986), 172–244. Documents cited here whose collection numbers begin with BM or NCBT are not included in the Brinkman and Kennedy catalogs.
The following is a list of the Babylonian documents which contain hardship formulae mentioned in this chapter (along with the publications in which there is a copy and/or translation of the documents).
H.2. Babylon, 25 Ab, Year 3 of Mushezib-Marduk, Brinkman, *JCS* 25 (1973), 93.

include dates that suggest they were composed in times of siege.[5] Some of these contain formulae explicitly indicating the suffering that prevailed at the time of the recorded deals. Oppenheim devoted a systematic and

---

J.24. Dilbat, 29 Nisan, Year 23 of Ashurbanipal, McEwan, OECT 10 399.
K.119. Babylon, 13 Marheshwan, Year 18 of Shamash-shum-ukin, San Nicolò, BR 8/7, no. 19.
K.128. Babylon, 24 Tammuz, Year 19 of Shamash-shum-ukin, Frame, *JCS* 51 (1999), 101–105.
K.132. Babylon, 5 Tebeth, Year 19 of Shamash-shum-ukin, unpublished (courtesy of Prof. J. A. Brinkman).
K.133. Babylon, 9 Tebeth, Year 19 of Shamash-shum-ukin, Pinches, *JTVI* 26 (1893), 163–165.
K.139. Babylon 22 Nisan, Year 20 of Shamash-shum-ukin, unpublished (courtesy of Prof. J. A. Brinkman).
K.140. …, 29 Iyyar, Year [20] of Shamash-shum-ukin, San Nicolò, BR 8/7, no. 20.
K.153. Babylon, 20 Sivan, Year … of Shamash-shum-ukin, Weidner, *AfO* 16 (1952–1953), 37, no. 2.
N.3. Babylon, accession year of Sin-shum-lishir, McEwan, OECT 10 400:13.
O.17. Nippur(?), during the rule of Ashur-etel-ilani or Sin-shar-ishkun in Babylonia, San Nicolò, BR 8/7, no. 63:24–26; see below, p. 148, n. 94.
O.23. Nippur, 6 Sivan, Year 3 of Sin-shar-ishkun, Oppenheim, *Iraq* 17 (1955), 87.
O.24. Nippur, 13 Sivan, Year 3 of Sin-shar-ishkun, Oppenheim, *ibid.*, 88.
O.25. Nippur, 14 Sivan, Year 3 of Sin-shar-ishkun, Oppenheim, *ibid.*, 88.
O.26. Nippur, 18 Sivan, Year 3 of Sin-shar-ishkun, Oppenheim, *ibid.*, 87.
O.27. Nippur, 10 Tammuz, Year 3 of Sin-shar-ishkun, Oppenheim, *ibid.*, 87.
O.30. Nippur, … Elul, Year 3 of Sin-shar-ishkun, Oppenheim, *ibid.*, 87–88.
O.33. Nippur, 10 …, Year 3 of Sin-shar-ishkun, Oppenheim, *ibid.*, 89.
O.35a. Uruk, 4 Marheshwan, Year 5 of Sin-shar-ishkun, von Weiher, *Spätbabylonische Texte aus Uruk* II, no. 57:42–59.
O.38. Uruk, 16 Elul, Year 6 of Sin-shar-ishkun, Hunger, *BaM* 5 (1970), 217–219, no. 12.
O.39. Uruk, 23 Tishri, Year 6 of Sin-shar-ishkun, Hunger, *ibid.*, 215–217, no. 11.
O.40. Nippur, 1 Marheshwan, Year 6 of Sin-shar-ishkun, San Nicolò, BR 8/7, no. 68.
O.41. Nippur, 21 Marheshwan, Year 6 of Sin-shar-ishkun, San Nicolò, *ibid.*, no. 69.
O.42. Uruk, 6 Shebat, Year 6 of Sin-shar-ishkun, Hunger, *ibid.*, 214–215, no. 10.
O.44. Uruk, 9 Ab, Year 7 of Sin-shar-ishkun, Walker, *AfO* 24 (1973), 125.
O.45. Uruk, 12 Tebeth, Year 7 of Sin-shar-ishkun, San Nicolò, BR 8/7, no. 71.
O.47. Nippur, 28+ Sivan, Year 3(?+) of Sin-shar-ishkun, Oppenheim, *Iraq* 17 (1955), 89.
O.54. Nippur, 11..., Year [3?] of Sin-shar-ishkun, Oppenheim, *ibid.*, 89.
P.1. Uruk, 2 Kislev, "Year of closure of the gate," Hunger, *BaM* 5 (1970), 210–212, no. 8.

comprehensive discussion to these documents, coining the term "siege documents" to describe them.[6] There are legal documents from Emar (Tell Ashtata, near Meskeneh, on the Middle Euphrates) from the second half of the 2nd millennium BC, which also contain hardship formulae similar to those found in the Babylonian siege documents. It is clear from the Emar documents that the existence of siege and hardship documents was not limited to Babylonia of the 7th century BC; rather, this phenomenon is evident over at least five hundred years. The type of transactions described in these documents is similar in many respects to those known to us from Babylonia, and all of them can enrich our knowledge of the realities of the hardship caused by famine.

For convenience in the examination of the legal purpose and economic significance of the siege documents, we begin by enumerating the characteristics of the hardship formulae in the two groups of documents, of Babylonia and of Emar. Oppenheim described the Babylonian siege documents in detail (see reference in n. 6); consequently, our discussion is limited to matters not discussed in his research, or about which we disagree with his analysis. The study of the Emar documents is still in its infancy, so the content of documents pertinent to the topic is elaborated briefly.

P.2. Uruk, 16 Kislev, "Year of closure of the gate," Hunger, *ibid.*, 208–210, no. 7.
P.3. Uruk, 16 Marheshwan, "Year 4, closure of the gate," Hunger, *ibid.*, 212–213, no. 9.
P.4. Uruk, 8 Shebat, "Year 4, closure of the gate," Hunger, *ibid.*, 235, no. 21.
P.5. Uruk, 8 Shebat(?), "Year 4, closure of the gate," Hunger, *ibid.*, 235–236, no. 22.
P.6. Uruk, 13 Marheshwan, "Year 5, closure of the gate," Hunger, *ibid.*, 223–224, no. 16.
P.7. Uruk, 6 Shebat, "Year 6, closure of the gate," Hunger, *ibid.*, 228–229, no. 17P.
P.8. Uruk, ..., "Year of clo[sure of the gate," Hunger, *ibid.*, 222–223, no. 15.
T.5.6. Uruk, 16 Ab, Year 5 of Nabopolassar, Scheil, *RT* 36 (1914), 191–192.
BM 114354. Uruk, 21 Shebat (name of the king not cited), Da Riva, *AuOr* 20 (2002), 249–251.
NCBT 511 (Uruk), "… Elul, during the closure of the gate," Beaulieu, *BaM* 28 (1997), 390.
NCBT 4904 (Uruk), 14 Marheshwan, "Year 4, hostilities in the land," Beaulieu, *op. cit.*, 388. On the significance of this date formula, see below, pp. 119–120.

6 Oppenheim, *Iraq* 17 (1955), 69–89.

II. THE BABYLONIAN SIEGE DOCUMENTS

The hardship formulae found in the Babylonian siege documents contain the following characteristics:

1. *A state of siege*: (a) a state of siege appears explicitly in Documents H.2:2; K.133:18; 139:54–55; 140:43–44; O.30:2; (b) the expression, "during the closure of the gate," with additional characteristics of the hardship formulae, appears in Documents H.2:6–7; J.24:5; O.30:2–3; 33:10–11; 47:4–5; (c) "at the (time)/year of the closure of the gate," without additional characteristics of the hardship formulae, appears in Documents N.3:13; BM 114354:32; NCBT 511:6; O.35a:59; 38:35; 39:34–35; 42:40; 44 rev. 16; 45:16; P.1:33–34; 2:31–32; 3:38; 4:18; 5:17; 7 iv 34; 8 L.E. 3'(?); compare further 82-3-23,3959 (BM 52925);[7] NCBT 4904:8–9; (d) an obligation to repay a loan "at the opening of the gate" appears in Documents O.17:26; 40:3–4; 41:3–4.[8]

2. *The notation "famine (and hardship) in the land"* appears in Documents H.2:2–3; K.119:62–63; 128:2; 132:2; 133:19–21; 139:56–57; 140:44.

3. *High prices of grain* are recorded in Documents H.2:5; K.132:3; 139:57; 140:44–45; O.30:3; 33:11; 47:5–6; compare J.24:4–5.[9]

4. The statement that a given transaction was made when *parents sold their children* appears in Documents K.153:6; O.30:4–5.

7 On this document, see below, p. 128, n. 39.

8 A document from Uruk details the obligation of a woman to return some beams, 24 cubits in length, "at the opening of the gate" (*ina patê bābi*); cf. Stolper, *BaM* 21 (1990), 571-572. The date recorded in this document is 10 Marheshwan, year 17 of Darius, " king of the lands" – that is, Darius II (i.e., the date is 15 November, 407 BC). We have no indication of a siege of Uruk at that time, and the political circumstances reflected in the document are unclear.

9 See J.24:5–6: *ki-i* (1?) MA.NA 1 GÍN KÙ.BABBAR *ina Bar-sip*.KI; "When a mina (was sold) for a shekel of silver in Borsippa" (on the omission of the kind of goods meant, almost certainly barley, compare below, p. 124, n. 26). While in the formulae recorded below the goods are given in units of volume (seah, *qû*), this document refers to units of weight. Although line 36 states that the document was written in Dilbat, it should be noted that it records the high price that was customary in Borsippa.

5. *Explicit literary motifs* may be detected in Documents H.2:3-5, 7-9; K.119:63-64.

1. A State of Siege

There are legal documents from Uruk containing the phrase "at the (time)/year of the closure of the gate" (*ina edil bābi*) on its own, without any further specification. Some of these explicitly mention Years 5, 6, and 7 of King Sin-shar-ishkun of Assyria[10] and Year 5 of King Nabopolassar of Babylonia.[11] Unlike the characteristics "famine and hardship in the land," high prices of grain, the selling of children by their parents and other explicit literary motifs, that may appear at the beginning of a contract (H.2; K.128, 132, 153; O.30), inside it (O.33, 47) or at its end (K.119, 133, 139, 140), the phrase "at the closure of the gate" always appears at the end of the document, as an integral part of the date. This phrase appears not only following specific mention of the year of the reign of the king, but also in the form "Year 4 (without citing the name of the king), closure of the gate,"[12] "year 5, closure of the gate,"[13] "year 6, closure of the gate,"[14] or even, simply "the year of the closure of the gate."[15]

10 Year 5 of Sin-shar-ishkun: O.35a; Year 6: O.38, 39, 42; Year 7: O.44, 45.

11 T.5.6.

12 P.3-5: MU 4 KAM *edil bābi*.

13 P.6: MU 5 [KAM] *edil bābi*.

14 P.7: MU 6 KAM *edil bābi;* NCBT 511:6 MU 6 KAM *ina edil bābi*.

15 P.1, 2, 8(?): MU (var. MU.AN.NA) *ša edil bābi*. In O.42, in which the formula "Year 6 of Sin-shar-ishkun, king of Assyria, at the closure of the gate" appears, we find the names of Nabû-damiq [lú]*šākin ṭēmi* and of Gimillu [lú]*šatammu Eanna*. These two officials are also mentioned in P.1 and 2, in which the formula "the year of the closure of the gate" appears without elaboration, while P.3 includes the formula "Year 4, closure of the gate." Documents from Uruk from the beginning of Nabopolassar's reign include the names of other officials who fulfilled the same functions: Anu-aha-iddin [lú]*šākin ṭēmi* (T.1.3, T.1.7, T.1.13, Year 1) and Kudurru [lú]*šatammu Eanna* (T.1.7, T.1.13, Year 1; T.5.22, Year 5; T.7.16, Year 7); cf. Kümmel, *Familie, Beruf und Amt im spätbabylonischen Uruk,* 139, 141-142. Prosopographic examination suggests that "Year 6, closure of the gate" in P.7 refers to the 6th year of Sin-shar-ishkun's reign (based on the names recorded in this document, those of the scribe and four of the five witnesses, which are identical to the names of the scribe and witnesses in O.38, 39, 42, that include the date "Year 6 of Sin-shar-ishkun, king of Assyria." For arguments against this identification, see Zawadzki [below, p. 122, n. 22], 181-182.

It seems unlikely that the number indicating the year in date formulae which do not mention the name of the king signifies the number of years since the beginning of the siege, as this would entail, for example, that Uruk was besieged for six years, based on Document P.7. It is also doubtful to conclude, based on the dates given in P.3 ("16th Marheshwan, Year 4, closure of the gate"), P.7 ("6 Shebat, Year 6, closure of the gate") and P.4–6 that were written between them, that Uruk was subject to a siege that lasted for 26 months at least. It seems, therefore, more likely that the number refers to the year of the royal reign (although it is not always easy to determine which king or kings are referred to in the formulae).

Closure and opening of the gates of the city were generally controlled by the defenders within and not by the besiegers (since the gate opened inwards and its bolt was *inside* the wall). Hence, it seems reasonable to assume that the expressions *ina edil bābi* ("at the closing of the gate") and *ina patê bābi* ("at the opening of the gate") were not necessarily limited to a *military* sense indicating an actual siege, [16] but were also used in a

---

A number of conclusions follow from this examination: (a) P.1–3 and 7 are from the reign of Sin-shar-ishkun; (b) there were two periods of "closure of the gate" in Uruk: the first recorded in documents from years 5–7 of Sin-shar-ishkun (O.35a, 38, 39, 42, 45; P.1–7, [8]) – and evidently also from Year 4 (compare P.3) – and the second noted in T.5.6, from year 5 of Nabopolassar. These conclusions must be taken into account in reconstructing the chronology and balance of power between Babylonia and Assyria in general, and in Uruk in particular, in the years 630–618 BC. On proposals for such reconstruction and on the difficulties involved, see Brinkman, *Prelude to Empire,* 109 n. 548; Beaulieu, *BaM* 28 (1997), 367–393; and cf. Zawadzki, *The Fall of Assyria,* esp. 37–41, 149.

One may postulate, as a somewhat peculiar alternative, that the officials in question were removed from their offices and then reinstated several times, as control of Uruk changed; this view is proposed, for example, by Zawadzki (*Folia Orientalia* 20 [1979], 184). Under the assumption that the administration in the Babylonian cities was continuous (and that officials did not change at times of transition of political rule), it is impossible to accept synchronization between Year 5 of Sin-shar-ishkun and Year 2 of Nabopolassar, as suggested by von Weiher, *Spätbabylonische Texte aus Uruk,* II. 219.

16 From document BM 52925 // 82-3-23, 3363, which was written in Babylon after it fell to Ashurbanipal, one gains the impression that the expression *ina edil bābi* refers to a state of siege and not to ordinary riots or uprising (see below, p. 128, n. 39).

political sense (i.e., the closing of the city by decision of its inhabitants, rejecting the government, or in an uprising).[17]

Periods in which the central Assyrian regime was challenged, and the eponym of the new year was not known, are marked by the date formula *limmu ša arki PN* ("the eponym year after so-and-so" (who was the eponym for the preceding year).[18] In Babylonia, it was customary to cite dates according to the regnal year even when the country was under Assyrian rule, but in some rare cases local scribes also used the Assyrian methods of year notation to note times of crisis.[19] It is therefore possible that the phrase *ina edil bābi* had a similar function (possibly even more efficient, as it could be used continuously for several years). According to this suggestion, the appearance of the phrase "Year 5 (or 6, or 7) of Sin-shar-ishkun, at the closure of the gate" in the documents from Uruk does not necessarily indicate that the city was under Assyrian control at that time, but only that during the struggle between Sin-Shar-ishkun and Nabopolassar for control of Uruk, the city scribes continued to record the year of the king's reign (in legal and economic documentation written in this inter-regnum period), even if he no longer ruled over the city in practice.[20]

Several years ago, a document from Uruk, NCBT 4904, was published, showing the date formula: "14 Marheshwan, Year 4 of hostilities in the land." According to the editor of the document, Beaulieu, this formula is not an alternative or political equivalent to the formula "closure of the

17 Compare Deut. 20:11; 2 Kgs. 15:16 ("opening of [the gate of] the city"); Streck, *Asb* 30 iii 105–108; OIP II 41 v 17–19 ("closing of the gate of the city"). Cf. above, n. 8.

18 For a collection of *limmu ša arki PN* date formulae from the Assyrian empire and for a discussion of their political significance, see Larsen, *RA* 68 (1974), 15–24; additional *limmu ša arki* were published by Dalley, *Abr-Naharain* 34 (1996–97), 75, 89.

19 For the use of local eponyms (different from those used by the Assyrians of the same period) and the formula *limmu ša arki PN* in documents from the 7th century BC but prior to the end of Assyrian rule in Babylonia, see Frame, *Babylonia 689–627 B.C.*: A *Political History,* 284–288.

20 The absence of the characteristics 2–5, indicating famine and hardship, in documents P.1–7, (8?), T.5.6, which include the formula "closure of the gate," lends support to this assumption. The documents in question were written by at least six scribes, indicating that the formula was accepted by many scribes and should not be seen as the idiosyncratic practice of a single scribe.

gate" noted above, that is attached to the notations for years 4, 5 and 6 (i.e., of the reigns of king/s whose name/s are not cited). According to Beaulieu, the formula in this document refers to *the fourth year of the hostilities,* which followed after three years of struggle between Sin-shar-ishkun of Assyria and Nabopolassar of Babylonia for control of Uruk, during the course of which the city went back and forth between two powers (in other words, this is a reference to years 623/2 – 621/20, i.e., years 3–5 of Nabopolassar = years 5–7 of Sin-shar-ishkun). This suggestion requires further documentation in light of our current information.

The economic significance of the state of "closure of the gate" may be gleaned from an examination of the nature of the documented transactions in which this expression appears. Following a study of 22 documents from Uruk found in the archives of Nabu-ushallim, Zawadzki concluded that these reflected sharp fluctuations in the prices of prebendary rights or income in kind from temples (*isqu*) that had been sold or mortgaged in exchange for loans near the end of the period of the Assyrian Empire.[21] In his view, the volume of such incomes increased (by a factor of as much as four) during the reign of King Sin-shar-ishkun of Assyria – a period of harsh and protracted struggle for the control over the cities of Babylonia – during which Uruk was in a state of siege and insecurity. The prices dropped significantly in the latter part of Nabopolassar's reign, once the political situation had become stabilized, although they never returned to the low level which had been the standard during the first seventy years of the 7th century BC.[22]

21 These documents were published by Hunger, *BaM* 5 (1970), 193–304. For *šattu 4^{kam} nukurti ina māti*, see Beaulieu, *BaM* 28 (1997), 367 ff., esp. 388.

22 See Zawadzki, *Folia Orientalia* 20 (1979), 175–184. Although Zawadzki's statement seems correct, one cannot ignore several problems in his analysis of the finds:
(a) Zawadzki ignores the fact that the prebends (*isqu*) enumerated in these documents refer to different temples and discusses them all as if they were equivalent in composition or quantity. However, these temples probably differed from one another in size and in the scope and times of ritual activity, differences which would have been reflected in the expected income of the groups connected with worship in them. (On the fixed benefits of various professions in the Babylonian temples during the Neo-Babylonian period that were paid in kind rather than in silver or in other forms of payment, see *in extenso* in MacGinnis, *AfO* 38–39 (1991–92), 74ff., and references therein.)

It should also be noted that, contrary to expectations, interest levels evident in the Babylonian documents recording loans that are marked by "closure of the gate" did not deviate significantly from interest levels in peace time.[23]

2. "Famine (and hardship) in the land"

One of the indications of food shortage appears in Document K.128 (written in Babylon a year after the beginning of Ashurbanipal's siege on that city), which notes that payment for the sale of a female slave was given in money (2 + x shekels) and in barley (60 *qû*).

The expression *sunqu ina māta šakinma,* "there is famine in the land," also appears in YOS VI 154:6–7, written in Uruk on 28 Adar, year 11 of Nabonidus of Babylonia.[24] According to this document, a woman whose husband had died, leaving her with two small children, gave her children to the Eanna temple. In other documents of the same year, there is no evidence of famine in Babylonia in general, or specifically in Uruk. It would therefore seem that the formula "there is famine in the land" is used in this particular document to indicate the hardship suffered by that specific woman and not that of the general public. It other words, this form is used as a technical (legal) formula that does not necessarily signal anything about the overall situation.

(b) Most of the income listed in the archives of Nabu-ushallim are of the type of *nuḫatimmūtu* (that is, derived from the franchise of baking or selling baked goods), but they also include some of the type of *sirāšûtu* (that is, derived from a franchise for production or sale of beer; K.165) and *mubannûtu* (derived from a franchise for the preparation or sale of other foodstuff, whose nature is unclear; P.1). Therefore, there is no certainty that the levels of income involved in all of these functions were the same.

Examination of the data in light of these reservations reduces by half the quantity of data upon which Zawadzki's conclusions may properly be based and the increase in prices of the types of income discussed.

23 Documents which record interest-free loans should evidently be considered a "charitable deed"; on this, see below, pp. 147–151.

24 For this document, see Dougherty, *The shirkûtu of Babylonian Deities*, 33.

3. High Prices of Grain

The prices of grain found in the Babylonian hardship formula are as follows:

"The price of two (seah) of barley in Babylon (was a shekel)" (K.132:3).[25]

"A seah (of barley) for a shekel of silver" (O.30:3; 33:11; 47:5–6, all these documents are from Nippur).[26]

"Three *qû* of wheat were purchased for a shekel of silver *in secret*" (see below, p. 134, n. 54; K.140:44–45, probably from Babylon).

"The price of three *qû* of barley (was) a shekel of silver" (K.139:57, Babylon).

"Two *qû* (of barley) for a shekel of silver" (H.2:5, Babylon).

The regular price for barley in Assyria and in Babylonia in the 8th and 7th centuries BC was one shekel of silver for a *homer,* with fluctuations of as much as up to 50% to 100%, depending on season and distance from the grain fields.[27] The prices noted in the hardship formulae are, therefore, 10

25 This fragmented formula does not cite any price; see next note.

26 See O.30:3: [*xxxx* K]I.LAM *šá* BÁN.TA.ÀM ŠE.BAR; O.33:11: *ma-ḫi-ri ša* BÁN.TA.ÀM; O.47:5–6: KI.LAM *šá* BÁN.TA.ÀM ŠE.BAR. The formulae of the price of barley in these three documents (written by at least two different scribes) are incomplete, probably because the formula was so familiar that the truncated form was sufficient; compare Deller (below, n. 27), no. 6 (from Sultantepe): S.U. 51/36:8–9: *ina* MU.AN.[NA] *šá* BANMIN *ana* 1 GÍN (the kind of merchandise and the metal in which it was paid are not given); a chronicle-like passage of Arik-den-ili, RIMA 1 126:18', 127:27', concerning the price of grain: 1 (var. 2) MA.NA.TA.ÀM ŠE *iššikin* (the unit of measurement and the metal used for payment are not mentioned); similarly, in the Panammua Inscription (*KAI* 215), line 6: "And a פרס cost a shekel […] and a אסנב [of oil?] cost a shekel" (פרס, cf., Akkadian *parīsu*, "half a kor" and אסנב, cf., Akkadian *šinipu*, "two thirds") are units of volume; the type of merchandise is not mentioned in the former passage, and the metal in which payment was weighed, probably silver, is not stated in either passage). Compare: "What is meant by 'stricken by drought'?… Rav Hannina said: If a seah is sold for a sela, and is available, that is drought [בצורתא], for four, and is unavailable, that is famine [כפנא] (*b. Taʿanit* 19a–b).

27 See Deller, *Orientalia* 33 (1964), 260–261, which also contains a transcription and translation of Letter NL 52 = ND 2355=CTN V, p. 193, reflecting the changes in the price of barley in accordance with the cost of transportation. On the prices of grain (mostly in the 6th and 5th centuries BC) and their fluctuations, cf. Meissner, *Warenpreise in Babylonien*, 5–6; Dubberstein, *AJSL* 56 (1939), 26–27. On seasonal fluctuations in the

to 50 times higher than the regular rates.[28]

Interesting parallels to the high prices recorded in the Babylonian hardship formulae are found in post-canonical Assyrian documents (i.e., written between 648–612 BC). These documents, dealing with the sale of property and female slaves, note that transactions were conducted "in the year in which the price of barley was *x* shekels of silver/minas of copper (URUDU) for *y* seah (Akkadian *sūtu*)/*qû*," as follows:[29]

"One mina of copper = 2 seah + 4 *qû*" (Deller, no. 2). Sale of a house (Assur).

"One mina of copper = 2 seah" (Deller, no. 5). Sale of a house (Assur).

"One mina of copper = […]" (Deller, no. 3). Sale of land and a house (Assur).

"One mina [of copper] = […]" (Deller, no. 4). The designation of the goods sold is not extant (Assur).

"One shekel of silver = 2 seah" (Deller, no. 6). Sale of a slave-girl for 4 silver shekels (Sultantepe).

"One shekel of silver = 1 seah" (Deller, no. 1). Sale of a slave girl for 20 shekels of silver (Assur).

"One shekel of silver = 1 seah" (*NATAPA* I, no. 41). Sale of a slave-girl for half a min[a] of silver (Assur).

"One shekel of silver = 1 seah" (*NATAPA* I, no. 43). Sale of a slave-girl for 10 shekels of silver (Assur).

---

price of barley in Babylonia, based on Old Babylonian documents, see Farber, *JESHO* 21 (1978), 17-21. From Babylonian astronomical diaries which also contain data on the prices of various foodstuffs, including barley, it follows that frequent significant fluctuations (as often as two or three times a month) were common in Babylonia. See Sachs and Hunger, *Astronomical Diaries and Related Texts from Babylonia*, I, 62–64, no. -418 obv. 4, 8-9, 15, 20, rev. 6'; 148-150, no. -346 rev. 11-12, 31-32, and many additional examples therein; and Vargyas, *A History of Babylonian Prices in the First Millennium BC*, 89-108, 120–127; Slotsky, *The Burse of Babylon*, 104–106.

28 This calculation is based on the ratio of 1 *imēru* (homer) = 100 *qû* (184 liter), which was dominant in Assyria during the Neo-Assyrian period; see Postgate, *Fifty Neo-Assyrian Legal Documents*, 67. On the same ratio in Neo-Babylonian documents, see Powell, *AOF* 17 (1990), 94.

29 Deller, *ibid.*, 257–259; Postgate, *ibid.*, 91–92; Fales and Jacob-Rost, *SAAB* 5 (1991), 89–90, 94–95. The documents are arranged in ascending order of the price of barley; the places in which the documents were discovered are given in parentheses.

"One shekel of silver = 1 seah" (Deller, no. 7). Adoption of a 15-year-old girl (Assur).[30]

"One [shekel of silver] = 1 seah" (Deller, no. 8). Sale of a slave for a mina of silver (Tell Halaf = Gozan).

"1½ mina of copper = 1 *qû*" (*NALD,* no. 8). Sale of a slave-girl for 20 minas of copper (Calah).[31]

The high cost of barley in the last document is concealed in a longer passage that may indicate the reason for this high price: *NALD* 8:15–17: *ina* MU.AN.NA *la ba-ši-ti a-ki 1 qû* ŠE.PAD.MEŠ 1½ MA.NA URUDU.MEŠ *ta-lak-u-ni* GEMÉ TI-*at*; that is, "the slave girl was taken [i.e., acquired, sold] in a non-existent year, when one *qû* of barley went (i.e., was sold) for 1½ mina of copper."[32]

The meaning of the expression "a non-existent year" is explained in document CTN II 15 (from Calah, 791 BC, sale of real-estate); in the concluding passage (after the list of witnesses and the date), it reads (line 52): *i-na* M[U.A].NA *uk-li la ba-ši-ti*; "in a non-existent food-year," that is, in a year of famine.[33]

30 The documents Deller, no. 1, and NATAPA I 41, 43, were all written in the same year: *limmu ša arki* ^{md}*Nabû-šarra-uṣur*.

31 Among the documents whose opening formula mentions a high price for barley is the Assyrian document SAA VI 285. In lines 3'-4' of this document, the editors read: [*ina*] ⌜MU⌝.AN.NA [*ša ina* 1 GÍN KUG.UD 6 BÁ]N ⌜ŠE.PAD⌝.MEŠ ⌜*tal*⌝*l-lik-u-ni* – "[In] the year [in which] [six se]ah of barley were sold (lit., went for) for a shekel of silver." The number of seahs mentioned in line 4' is surprising, as a price of 6 seah for a shekel of silver does not seem so unusual as to justify its being mentioned in the introductory formula of a legal document.

32 On the terms used in buying and selling, compare *tos. Avodah Zarah* 4.4: "A person may not leave the land (of Israel) unless wheat is sold (הולכין, "going") at two seah for a sela. Rabbi Shimon said, to what does this refer? At a time when he cannot find any to buy. But when he can find some to buy, even if at a seah for a sela, he should not leave."

33 For the formula under discussion, see Postgate, CTN II, 45–46. An Aramaic papyrus in the Demotic script (Pap. Amherst Egyptian 63), includes a description of a period of drought and food shortage as י(ו)מן דלהוון שנן דל(א)פקו, that is, "days which had not been, years which had not been spent(?)/turned(?)"; see Steiner and Nims, *RB* 92 (1985), 70–71. It follows from this expression that the word *uklu* in the expression "in a non-existent food-year" in CTN II 15:52, is an explanatory phrase not included in the original expression.

The Assyrian formulae brought here state that they were written in a *year* when the price of barley was high or when there was no food at all, but not during "the closure of the gate." Therefore, it can be concluded that they allude to years of drought or some other natural disaster, and not necessarily to times of siege.[34]

The biblical story of the siege on Samaria imposed by King Ben-hadad of Damascus, includes a description of "a great famine" during a period of siege: "until an ass's head was sold for eighty shekels of silver, and a quarter *qab* (comparable to Akkadian *qû*) of dove's dung (חריינים; *qeri* דביונים) for five shekels of silver" (2 Kgs. 6:25). The sharp change of this stark reality is described in the words of the prophet Elisha, "Tomorrow at this time a seah measure of fine meal shall be sold for a shekel, and two measures of barley for a shekel, at the gate of Samaria" (*ibid.* 7:1, and cf. *ibid.* vv. 16, 18). Ostensibly, there is nothing remarkable in the latter prices, which are still significantly higher than the regular price; but it would seem that the text does not in fact refer to the regular prices of grain, but rather to the prices which will be in effect in the marketplace immediately upon cessation of the siege (hence the expression, "tomorrow at this time"). The prices certainly continued to decrease over the following days, but even upon cessation of the siege, they were already far lower than they had been during the siege.

4. Parents Sell Their Children

Eight of the nine sale documents extant in the archives of Ninurta-uballit in Nippur (dating from year 3 of Sin-shar-ishkun's reign) are concerned with the sale of girls (O.23–27, 30, 47, 54), while only one of them records the sale of a male child (O.33).[35] In two of these documents, the reasons for the sale of the children are given as: (a) "to sustain them" (i.e., to feed them

34 The dates of the post-canonical Assyrian documents recorded above are extant in only six cases, and these give four different eponym years, 643, 626, 625 and 612 BC according to the system of dating used by Falkner, *AfO* 17 (1954–56), 100–120.

35 On the sale of female children due to famine (not necessarily in connection with siege), compare: *ACh Sin* XXV 7–8 (p. 24): *ubbuṭu ibbaššīma iššakkan nišū mārātišina ana kaspi ipaššarā*, "There will be famine, people will sell their daughters for money."

and to keep them alive);[36] in a fragmented passage in O.30:14, we read: [*ki l*]*a še-bu-ú i-maḫ*(?)-*x;* that is, "[if she (the girl sold) is no]t satiated, then he….," stating the obligation of the purchaser as an explicit condition of the transaction; (b) so that the parents can buy food with the proceeds of the sale (6 shekels) and eat.[37, 38] On the literary nature of this characteristic as a whole, see below, pp. 129ff.

The expected fate of children sold during a siege may be reflected in a complaint, lodged by a woman against her (former?) husband, preserved in a fragmented document from Babylon:[39] "My son is gone (*mārua iānu*), and I have received nothing in your home; and at the closing of the gate you sold for money (the slave) Bel-tallak, my dowry which my father gave. You and your brothers... and the members of your house..." The phrase

36 O.27:4, 30:8. Compare also: YOS VI 154:9 *bul-liṭ-a-...* (on this document, see above, p. 123, n. 24). On the significance of *bulluṭu* as "to keep alive, to sustain," see below, p. 144, n. 83.

37 O.27:5–6, 30:9. These documents were written by two different scribes. (My thanks to Prof. J. A. Brinkman for making available to me the names of the scribes of the Ninurta-uballit archives preserved in the Cuneiform Collection at the Oriental Institute of the University of Chicago.)

38 On the connection between famine and the sale of children, compare Leichty, *Izbu* XVII 73' (p. 170): *mātum innešši sunqu dannu ina māti ibaššīma nišū mārāšina ana kaspi ipaššarā,* "Riots will break out in the land, there (will be) a great famine in the land, people will sell their children for money."

39 This document, published by Roth, *AfO* 36–37 (1989–1990), 50, is parallel to 82-3-23, 3363 = Pinches, *RT* 19 (1897), 107–108, or, more likely, is the same document (the different numbering for the same document may be explained as the result of an error in numbering in Pinches' publication or in the catalog of the British Museum). In Roth's edition, Pinches' reading from line 5 on is corrected; rather than *ina e-dil bābi ana* ^{d}*Bēl tal-lak* – "At the closing of the gate you went to (the god) Bel," one should read here ^{md}*Bēl-tal-lak*, that is, the name of the slave who belonged to the complainant as part of the dowry she received from her father, and who was sold "at the closure of the gate." The new edition includes sections that are missing in Pinches' edition, from which it follows that the document is a copy of an earlier document and that it was written in Babylon. The date is given according to the year of the reign of Ashurbanipal (the number of the year is not extant), after the conquest of the city from Shamash-shum-ukin, from which it follows that "the closure of the gate" mentioned in the document was the period of rebellion against Ashurbanipal (compare Streck, *Asb* 30 iii 105–108), and perhaps also the period of the protracted siege of the city.

"my son is gone" (and not "my son is dead")[40] raises the possibility that the circumstances of the child's disappearance are connected with his being taken away from his mother at "the closure of the gate," and that he may have been sold and changed masters until it was no longer possible to trace him.

5. Explicit Literary Motifs

Oppenheim has noted the source of the phrase "Mother did not open door to the daughter" (K.119:63–64: *ummu ana mārti ul ipatti bābu),* and to a certain extent also that of the characteristic already mentioned, namely, the sale of children by their parents, in the *topos* known as *Fluchtzeitschilderung* (description of hardship), found in Sumerian literary works and reappearing in Akkadian works. He noted clear parallels in the following works:[41]

a) The Babylonian *Atraḫasis Epic* describes the creation of mankind and the proliferation of the human race until they disturb the tranquility of the gods. The latter decided to drastically reduce the number of humans, first by famine and then by flood. The epic describes the effect of this famine, increasing year by year, "When the fifth year arrived, daughter watched the mother's going in, but the mother would not open her door to the daughter. The daughter watched the scales (i.e., the weighing of the silver at the sale) of the mother, the mother watched the scales (at the sale) of the daughter."[42] When the sixth year arrived, they served up the daughter for dinner, they served up the son for food."[42]

b) *The Prophecy of Shulgi,* that belongs to the literary genre known as "apocalyptic prophecies" (in fact, these are pseudo-prophecies, written *post factum*), includes descriptions of future catastrophes predicted to occur in the reigns of various kings. The reign of one of these kings is described as:

40 Compare Gen 37:30: הילד איננו – "the lad is gone"; *ibid*, 42:13, והאחד איננו ... – "and one is no more."

41 Oppenheim, *Iraq* 17 (1955), 78–79.

42 Lambert and Millard, *Atra-ḫasis* 112–113 vi 7–10:
7 *5 šattu ina kašādi erēb ummi mārta idaggal*
8 *ummu ana mārte ul ipattu bābša*
9 *zibānīt ummi mārtu inaṭṭal*
10 *zibānīt mārte inaṭṭal ummu*

"that prince will experience misery, and will have no satisfaction. So long as he is king, fighting and warfare will not cease… people will sell their children for silver… mother will bar her door against daughter."[43]

c) A similar formula appears in the Sumerian *Lamentation over the Destruction of Ur*, older than the siege documents discussed here by a thousand years or more: "The mother left her daughter; the people groan. The father turned away from his son; the people groan."[44]

Other sources can be added to the sources mentioned above, such as:

d) The Babylonian omen series *Šumma izbu* includes the phrases: "There will be bad times. The mother will bar her door against her daughter; there will be no [sense of] brotherhood."[45]

e) The treaty of King Esarhaddon of Assyria with Median chieftains, enumerates the curses that will befall the chieftain who violates the treaty, among them the phrase (line 448): "mother will [bar her door] against daughter."[46]

The motif of the mother barring her door to her daughter is part of a broader picture, describing the dissolution of the very foundations of society and the disintegration of the family framework in times of siege. An elaboration of this phenomenon appears in the description of Ashurbanipal's siege against Babylon in his Prism C, "The mercy of the people had come to an end. The father on his son, the mother on her daughter, had no mercy. The man abandoned his wife, [the father] deserted his beloved son."[47]

These motifs, a mother barring the door to her daughter and parents selling their children, are only details reflecting a well-known phenomenon in Mesopotamian literature; scribes had an abundance of motifs and

43 Borger, *BiOr* 28 (1971), 15:10'–11', 15':
10' *nišū mārīšina ana kaspi ipaššarā*
15' *ummu eli mārti bābša iddil*

44 Kramer, *Lamentation over the Destruction of Ur*, 43:233–234; for a similar, albeit not identical, motif, see *ANET*, 459.

45 Leichty, *Izbu* I 50: *lumnu ibaššima ummu eli mārtiša bābša iddil aḫḫūtu ul ibašši.*

46 SAA II, 46. For the correct reading of this line, see Borger, *ZA* 54 (1961), 188.

47 ND 5406 ii 16–20: *rēme nišē iq[tâ] abu ana mārišu ummu ana mār[tiša] ul irašši rē[mu] eṭlu ḫiratsu unda[ššir abu] ētezib māru nāram [libbišu]* (Knudsen, *Iraq* 29, 1967, 56).

examples at their disposal, to be used in modular fashion in the various genres.[48] Literary parallels to another motif in the hardship formulae exemplify this phenomenon further: "The corpses of the people were scattered in the squares of Babylon, with none to bury them" (H.2:7-9).[49] The description of Babylon after its conquest by Ashurbanipal (at the end of the siege depicted in Documents K.119, 128, 133, 140, 153) reads: "The corpses of the people whom the plague laid low, and those who lost their lives through hunger and want – what was left of the feast of the dogs and swine[50] – whose members blocked the streets and filled the squares, I ordered to be removed from Babylon, Kutha and Sippar, and cast (outside) upon heaps" (Prism A of Ashurbanipal, iv 79-85).[51]

While the penetration of expressions and technical terms of daily life into literature is considered a normal and self-evident phenomenon, the penetration of literary phrases into legal documents, as is found in the siege documents mentioned above, is unusual and requires some explanation. A factor that may help to account for this phenomenon is the complex nature of Akkadian writing, which demanded extensive training of scribes

48 The migration of motifs among different genres in the literature of the ancient Near East, in general and in Mesopotamian literature in particular, is quite a common phenomenon, although it has not been comprehensively discussed in the research thus far. For examples from the Mesopotamian literature, the data in the following works will suffice: Parpola, in *Language, Literature, and History: Philological and Historical Studies Presented to E. Reiner*, 271-274; Livingstone, *Court Poetry and Literary Miscellanea*, xxii-xxiv; Hallo, in *Lingering Over Words: Studies in Ancient Near Eastern Literature in Honor of W. L. Moran*, 203-217.

49 *pagrē nišē ina la* [lú]*qēbiru muṣṣâ rebâti Bābili* (Courtesy Prof. J. A. Brinkman); note that, despite the bleak picture painted, the city of Babylon was able to withstand Sennacherib's siege for 15 months (Brinkman, *JCS* 25 [1973], 93-94). Compare Jer. 14:16: "And the people... shall be left lying in the streets of Jerusalem because of the famine and the sword, with none to bury them -- they, their wives, their sons, and their daughters..."

50 Feeding dogs, swine and the like with the flesh of human bodies is a literary motif used to indicate the punishment of sinners and criminals who were not buried after their execution.

51 Streck, *Asb.*, 38-40: *pagrē nišē ša Irra ušamqitu u ša ina sunqi bubūti iškunū napištu riḫīt ukulti kalbē šaḫê ša sūqē purrukū malû rebâti ultu qereb Bābili Kutû Sippar ušēṣīma attadi ana kamâti*; cf. Cogan and Tadmor, *Orientalia* 50 (1981), 231-232 (BM 134436 col. A 3'-11').

(compared to the simple alphabet which was used in the West and was available to all). Scribe training involved familiarity with different genres and literary works. Even a scribe whose professional activity was restricted to the writing of legal and administrative documents became acquainted with literary texts in the course of his studies, and occasionally displayed this knowledge in the documents he wrote.

A range of characteristic features is evident in the Babylonian hardship formulae, from the realistic to the literary: at one extreme is the most concrete, "the closure of the gate" (feature 1), while at the other extreme are explicitly literary motifs (feature 5). Where does the third feature, the high price of grain, fall within this range?

The wide variations in the price of grain noted in the hardship formulae in the Babylonian siege documents would appear to reflect actual fluctuations in the price of the most basic staple in Mesopotamia. However, this conclusion is doubtful. It would seem more accurate to evaluate the *hardship topos* expressed in the high prices of grain as a parallel to another *topos*, customarily referred to as the *abundance topos*, known from Mesopotamian royal inscriptions (King Sin-kashid of Uruk, 19th century BC; Shamshi-Adad I of Assyria, 18th century; Ashurbanipal of Assyria, 7th century and Nabonidus, the last king of Babylonia, mid-6th century), and for which there are parallels in Hittite hieroglyphic inscriptions as well.[52] In these royal inscriptions, abundance is expressed by noting particularly low prices, especially of basic staples, the most common of which were barley, oil and wool. In light of the data found in numerous contemporary documents of sale, it has already been established that the prices cited in the *abundance topos* in the royal inscriptions are exaggeratedly low compared to the economic reality predominant at the time of their writing. The clearly low prices found in the royal inscriptions do not indicate actual market prices during the reigns of the kings in question; rather, they reflect the creative imagination of the scribes who composed these inscriptions, each according to his own taste, provided that all these formulae noted

52 On the low prices given in the *abundance topos* in Mesopotamian and Hittite royal inscriptions, see Hawkins, in *Ancient Anatolia: Aspects of Change and Cultural Development*, 93–102.

particularly low prices. Just as the low prices in the *abundance topos* should not be viewed as realistic data, so should the high prices in the hardship formulae not be viewed as real. Accordingly, it is not surprising that the price of barley in Siege Document O.47, written in Nippur in Sivan of Year 3 of Sin-shar-ishkun, namely "one silver shekel for one seah of barley," is the same two months later, according to another document written in the same city (O.30), although it seems likely that the prices rose the longer the siege continued. The same holds true for the price of barley cited in the hardship formula in Document H.2 from Babylon, namely "one silver shekel equals one *qû* of barley," which reflects greatly inflated prices; however, it is known that the city continued to withstand the siege for another fifteen months before it fell into Sennacherib's hands.

Oppenheim found no mention of high prices of grain in the literary works mentioned above, and this may have been one of the reasons that led him to refrain from identifying the price of barley simply as part of the *hardship topos* in the documents in question. Yet, if we proceed in his footsteps, we will find evidence for the literary nature of the missing element. *The Curse of Akkad* describes the famine and shortage that descended upon the city. Lines 162–169 include a description of the desolation of the roads, along the rivers and in the grazing areas, from which the goats and cows had disappeared and where only bandits were to be found; the inhabitants of the cities were trapped behind the walls. Lines 171–183 read as follows:

> (For the first time) since cities were built and founded,
> The great agricultural tracts produced no grain,
> The inundated tracts produced no fish,
> The irrigated orchards produced neither syrup nor wine,
> The gathered clouds did not rain, the *mašgurum (*a therapeutic plant) did not grow,
> At that time, one-half *qû* of oil was sold for a shekel,
> One shekel bought only one-half *qû* of grain,
> One shekel bought only one-half mina worth of wool,
> One shekel's worth of fish filled only one seah (=10 *qû*)
> They were sold at such (prices) in the markets of all the cities (whose inhabitants were trapped therein)!

> He who slept on the roof died on the roof,
> He who slept in the house, had no burial,
> People were flailing at themselves from hunger.[53]

In the light of this passage, can we assume that the high price of barley (a staple in the diet of the average Mesopotamian) in the hardship formulae was merely a literary *topos* and not actual information (compare also the literary nature of the motif of purchasing barley at a high price *in secret:* K.140: 44–45)?[54]

53 Cf. Cooper, *The Curse of Agade*, 59.

54 The motif of selling in secrecy is intended to exacerbate the feeling of food shortage: it is impossible to buy grain "at the gate / in the marketplace," even at outrageously high prices, because the buyer (and not only the seller) does not want to make his purchase public knowledge, so that he will not need to share the grain purchased (at great cost) with acquaintances and relatives. In Deuteronomy 28:49–57 the motif of eating in secret is combined with that of cannibalism in connection with siege, but under circumstances far worse than those described in Document K.140: "The Lord will bring a nation against you from afar, from the end of the earth… a ruthless nation, that will show the old no regard and the young no mercy. It shall devour the offspring of your cattle and the produce of your soil… leaving you nothing of new grain, wine, or oil, of the calving of your herds or the lambing of your flocks… It shall shut you up in all your towns throughout your land, until every mighty, towering wall in which you trust comes down… And you shall eat your own issue, the flesh of your sons and daughters whom the Lord your God has given you, because of the desperate straits to which your enemy shall reduce you. He who is most tender and fastidious among you shall begrudge his brother and the wife of his bosom and the children he has spared to share with any of them the flesh of the children that he eats, because he has nothing else left as a result of the desperate straits to which your enemy shall reduce you in all your towns. And the most tender and dainty woman among you, who would never venture to set her foot on the ground, shall begrudge the husband of her bosom, and her son and her daughter, the afterbirth that issues from between her legs and the babies she bears; she shall eat them secretly, because of the desperate straits to which your enemy shall reduce you in your towns." (On Deut. 28:49–52, compare, with certain differences, Jer. 5:15–17; to Deut. 28:53[-57], compare Jer. 19:9.) A. B. Ehrlich, *Mikrā ki-Pheshutô (Die Schrift nach ihrem Wortlaut),* [Hebrew] vol. 1 (Berlin, 1899), 367, properly comments: "'In secret' – against her husband and her grown sons and daughters, that they not take their portion. And it does not use the word 'in secret' with regard to the man, because the man has greater strength and can eat in the presence of his wife and children without giving them any of his food without fearing them."

There is an apparent difficulty in speculations regarding the borrowed literary motif and its inclusion in Akkadian legal documents (written in the 7th century BC, as noted above): all sections of *The Curse of Akkad* that have been found thus far are written in Sumerian and not in Akkadian. A similar description ("a seah of barley was purchased for ...") is found in the *Epic of Naram-Sin*, in the context of a prophecy of the evils that Enlil will bring upon the enemies of Naram-Sin, evils that will include siege and acts of hostility between the cities and between people (lines 131–143).[55] However, this passage from the *Epic of Naram-Sin* appears only in a Late Assyrian version, and cannot be taken as evidence for the existence of an earlier Akkadian source for the formula of high prices. The chronological and linguistic distance between the two groups of sources discussed requires connecting links in order to confirm the assumption regarding the process proposed here. Such a link has been found in the Emar documents, in which hardship formulae were preserved (as the result of hostilities and famine). Therefore, we will now turn to a discussion of these formulae, and later note the nature of the transactions reflected in the documents in which they are recorded.

III. DOCUMENTS OF SIEGE AND HARDSHIP FROM EMAR[56]

As noted above, among the Emar documents from the second half of the 2nd millennium BC are legal documents which include elements characteristic of the Babylonian siege documents. These elements are: (1) notation of the situation of siege, (2) a formula stating that the transactions were

55 See Gurney, *AnSt* 5 (1955), 106.

56 While the Hebrew version of this book was already in advanced stages of preparation for publication, a paper was published by Zaccagnini, titled "War and Famine at Emar," *Orientalia* 64 (1995), 92–109. To my surprise, there was much similarity both in approach and in conclusions between this paper and my own discussion of the Emar documents. I therefore saw no need to change anything in the original version of the book. Readers of Zaccagnini's paper will find there several hardship documents from Emar not included in the present book, as well as discussion and expansion of several points that are only briefly discussed here.

conducted "in the year/time of hardship (and) hostilities" and (3) the high price of grain.

1. Explicit Evidence of Siege

a) *ASJ* 12 (1990), 190 no. 7: a document of ownership recording two activities relating to one person (Mashruhe, the diviner): (i) the purchase of two houses for the price of 600 shekels of silver from (the god) Ninurta and the elders of the city (the houses were confiscated from Baal-gamil son of Irib-Baal because he had sinned toward his master: ll. 1–20)[57]; (ii) the granting to Mashruhe of a large field (whose dimensions were 1 by 2 *ikû*) by King Pilsu-Dagan. The circumstances of this event are recorded in lines 29–37: "As the Hurrian troops surrounded the city wall of Emar, and the divination of Mashruhe, diviner (lú *bārû*) of the king and the city, came true, Pilsu-Dagan, the king, has given him this field as a present" (on the transaction as a whole, see ll. 21–40). At the top of the list of witnesses appears the name of [Pi]lsu-Dagan son of Baal-kabar[58]; the scribe is [Abi]-kapi.

b) *RPAE* 42: this is a "collected" text (*Sammeltext*) containing copies of three votive inscriptions (each noting the dedication of a golden cup, weighing 30 shekels, to Baal): two of these are from the reign of (King) Pilsu-Dagan son of Baal-kabar of Emar, and one is from the reign of his son Elli. The scribe of the three original inscriptions is Ea-mudammiq. Lines 8–19 state: "(In the reign of)[59] Pilsu-Dagan son of Baal-kabar, king of

57 A hardship formula is not recorded in this part of the document. Since one of the two legal transactions recorded in the document is connected to the Hurrian siege on Emar, one might well ask whether the first activity is also related to that same time-period. If this is indeed the case, one may assume that Baal-gamil's sin toward his master relates to the same incident, namely, to a violation of the loyalty to the king and to collaborating with the enemy.

58 A list of the witnesses appears at the end of the document and relates to both transactions. We may therefore surmise that both were carried out at the same time. On a similar phenomenon, see below, p. 139, n. 67.

59 The circumstances of the dedication of the cup are recorded in the document in a parenthetical remark, without an opening word such as *inūma* (compare *ASJ* 12 [1990] 190 7:29, cited above). In our translation, this word is needed to understand the context.

Emar, the king of the Hurrians besieged[60] Emar. (But) Pilsu-Dagan raised his eyes to Baal and Baal gave him (a sign? by the flight of a bird?) as he wished (*eg-re-ti* MUŠEN GI ša ŠÀ-šú). Then he (Baal) struck the warriors (of the king of the Hurrians) (who) were both in the city and on the wall, and saved Emar."[61]

c) *TSABR* 25: a document of enslavement of one person to another who, "in the year in which the TAR.PI army has surrounded the city[62] and one *qû* of barley (went) for one shekel of silver," sustained him by giving him twenty *parīsu* of barley, thereby saving him from starvation.

d) *TSABR* 44: a document of enslavement of a person who, "in the year of hardship, when the TAR.PI army had besieged Emar," (ll. 2–3) was unable to pay off his debt in the sum of twenty shekels of silver.

2 (a). Explicit Mention of Hostilities (War)

In the context of the enumeration and description of transactions, the following documents indicate that they were drawn up "in the year of hostilities," that is, in the explicit context of warlike activities:

a) *RPAE* 20:14: the purchase of a house "during the year of hardship (and) hostilities" ([*a-na* M]U-*tu*$_4$ KALA *nu-kúr-ti*). Lines 28–32 state

60 The text reads [uru]*E-mar i-la-min,* and is translated accordingly by Arnaud and other editors of the inscriptions (see below, n. 61) as "maltraitait la ville d'Emar." But would it not be better, despite the orthographic difficulty, to connect the verb *ilamin* with *lamû* ("to besiege") rather than with *lemēnu* ("to mistreat")? Compare, evidently in the context of the same event, *ASJ* 12 (1990), 190:29–30: *i-nu-ma* ERIM[meš] *Ḫur-ri* BÀD [uru]*E-mar*[ki] *il-mi,* "as the Hurrian troops surrounded the city wall of Emar."

61 For a corrected transcription of lines 8–19 and discussions of their contents, see Tsukimoto, *ASJ* 12 (1990) 191–192; Dietrich, *UF* 22 (1990), 33–35; Fales, in *Marchands, diplomates et empereurs,* 81–84. The two documents discussed here (*ASJ* 12 (1990), 190 no. 7; and *RPAE* 42) are not explicitly siege documents, because they do not have a formula indicating the hardship of the public or of the individual at the time the grant was executed.

62 The same term also appears in *TSABR* 44:2. The etymology of TAR/ṬAR-PI/WU and the identification of those referred to by the term in the Emar documents are not sufficiently clear. There are those who identify them with the army of the Hurrian king, whose siege of Emar is mentioned in other documents. For the different views on this question, see Vita, *AOF* 29 (2002), 121–122.

that the woman who sold the house abandoned her children "because of hardship (and) hostilities" (*a-na dan-na-ti nu-kúr-ti*).

b) *RPAE* 111:36–39: a sale document of a house at the end of which, following the list of witnesses, appears the formula "the years of hostilities (and) hardship, in which you sell children whom you love, be it a daughter or a son" (MU.ḪI.A KÚR-*kúr-ti* KALA.GA *i-na* DUMU.MEŠ-*ši a-šar ta-ra-am ta-na-din i-na* MÍ-*ti ú* NÍTA).

c) *RPAE* 256:10: this document speaks of the adoption of a male child and two female children after their father had died "in the year of hostilities" (*a-na* MU.KÁM *ša* KÚR.MEŠ *nu-ku-ur-ti*).

d-f) *ASJ* 12 (1990), 207 15:15; *AuOr* 5 (1987), 221, 6:15; *ET* 9:19: the purchase of fields that were sold "because of hardship (and) hostilities" (*a-na dan-na-ti nu-kú[r-ti]*; without using the expression "in the year of…").

g) *ASJ* 12 (1990), 209 16:14: the purchase of a field "at (the time of) hardship (and) hostilities" (*i-na dan-na-ti nu-kúr-ti;* without the expression, "in the year of").

h) RE 31:21 Purchase of land "at (the time of) hostilities, in the year of hardship" (*i-na nu-ku-ur-ti i-na* MU.⸢KAM⸣ KALA.GA).

j) RE 35:13 Purchase of a field "in the years of hostilities (and) hardship" (*i-na* MU.ḪI.A *nu-kúr-ti* KALA.G[A]).

k) A will which includes a transfer of ownership of a house and orchard (and possibly of other property).[63] After the list of witnesses and of the name of the scribe, appears the notation: "years of hostilities (and) hardship" (MU.ḪI.A *nu-kur-ti* KALA.GA). The significance of its inclusion in this document is unclear, since the execution of the terms of the will was not influenced by conditions at the time the document was written, but rather at an unforeseeable time in the future.

The list of witnesses in Document (c) begins with King Zu-Ashtarti, son of Baal-kabar king of Emar, the list of witnesses in Document (g) begins with King Pilsu-Dagan, who also appears at the beginning of the list in Document (a), albeit without the royal title. This suggests that there were at least two "periods of hardship and hostilities" in the history of Emar:

63 Dalley and Teissier, *Iraq* 54 (1992), 94.

one during the reign of Zu-Ashtarti and the other during the reign of Pilsu-Dagan.[64]

2 (b). The General Designation "In the year of hardship"[65]

a) The general designation, "in the year of hardship," whose meaning was certainly clear to every reader of the Emar documents, most likely refers to famine. In several documents, this expression appears in conjunction with the detailing of past or present circumstances in which the document was written: *RPAE* 83:3,[66] 86:4, 138:11, 24, 41,[67] 139:42, 149:38, 158:14, 162:10'; *TSABR* 52:5, 65:6, 74:7; *ASJ* 10 (1988), 160, Text C, line 8. Sigrist,

64 For a list of the kings of Emar and the chronological relationship between them, see Arnaud, *Syria* 52 (1975) 87–92; Skaist, *ZA* 88 (1998), 45–71. On the siege during the reign of Pilsu-Dagan and its circumstances, see Skaist, *ibid.*, 64–67; Vita, *ibid.*, 115–123. On additional hardship documents, other than documents (a) and (g), in which Pilsu-Dagan son of Baal-kabar is listed as a witness (although he is not mentioned there explicitly with the title of "king"), see *RPAE* 138,158. Elli, the son of Pilsu-Dagan, is listed as a witness in *RPAE* 139, which is also a document from "the year of hardship." See also *RPAE* 42, which mentions the siege on Emar in the days of Pilsu-Dagan, above, p. 136.

65 On the various forms of writing the formula *ina/ana šatti dannati*, see Tsukimoto, *ASJ* 10 (1988), 162. One may distinguish variations in the reference of hardship even in the headings of the same scribe. Thus, for example, the following documents, written by the scribe Dagali, include the following features: *RPAE* 162:9'-10': the date formula at the end of the document says: "The month of Baal [...], year of hardship [...]"; *TSABR* 57: the sale of a house "because of hardship"; *ASJ* 12 (1990), 209 16:14: "at (a time of) hardship and hostilities"; and likewise in the different forms of writing used by the scribe Ish-Dagan:
RPAE 138:11 *i-na* MU.1.KÁM KALA.GA
RPAE 138:41 *a-na* MU.1. KÁM KALA.GA
RPAE 138:24 *a-na* MU KALA.GA
RPAE 139:42 *i-na* MU KAL.LA.GI

66 For a corrected version of this line, see Tsukimoto, *ibid.*, 162.

67 *RPAE* 138 records a sequence of three transactions carried out by one person (Iṣṣur-Dagan, son of Baal-kabar, from the royal family), each of which is recorded as carried out "in a year of hardship." This refers to the purchase of a field (for the price of 20 silver shekels), a vineyard (50 shekels) and a house (60 shekels), bought from three different owners. The fact that all three transactions are recorded in a single document, which ends with a single list of witnesses (all from the royal family) suggests that all three were carried out together.

kinattūtu ša dārâti, No. 2:2; *AuOr* 5 (1987), p. 231 13:3; Arnaud, *SMEA* 30 (1992), 210–212, No. 9. The last document is an adoption contract wherein the adopter, who did not have sons, adopted a man and took him together with his wife and two sons into his home, in order to take care (*palāḫu*) of him (the adopter), and "[that he will pay his de]bt and [keep alive, sustain] his two daughters in [a] year of hardship" (*ina* MU.KÁM KALA.GA). *RPAE* 196 is a badly fragmented will, line 2' of which includes the words "[in a ye]ar of hardship." *TSABR* 57 speaks of the sale of a house "because of hardship" (line 18), without using the words "in the year/time of."

b) In several documents, this general expression appears in conjunction with a phrase referring to the high price of barley: *RPAE* 121; *TSABR* 74; *ASJ* 10 (1988), 166, Text E; 13 (1991), no. 37. Concerning these four documents, see the next page.

Years of famine resulting from drought were not an unusual phenomenon in the period in which the Emar documents were written.[68] It is therefore clear that the formula "in the year of hardship" cannot in itself serve as an indication of war and siege.[69] The probability of the formula being related to siege increases only if the document can be connected to other documents which make more explicit mention of the circumstances, primarily by prosopographic examination.

3. High Prices of Grain

(a)-(c) *ASJ* 10 (1988) p. 166, Text E, ll. 1–2; 13 (1991), p. 302 37:2–3; Sigrist, *kinattūtu ša dārāti*, No. 2:2–3: "In the year of hardship, when three

68 See, for example, Klengel *AOF* 1 (1974), 165–174.

69 Compare, for example, the formula *sunqu ina māta šakinma* (YOS VI 154:6–7) in a document from Uruk from the eleventh year of Nabonidus, whose connection to siege is doubtful: see above, p. 123. On the other hand, it is noteworthy that the absence of any explicit mention of siege in these documents does not necessarily indicate that one is speaking of ordinary famine and not of war-related famine. Thus, for example, we have extant documents from Babylon which include the formula "famine and hardship were in the land" without any mention of siege; only on the basis of the dates recorded in them can it be established that these were written during Ashurbanipal's battle against the city (see K.119, 128, 132).

qû (in the third document: three *qû* of barley) went for (lit, stood at) a shekel of silver."[70]

(d) *RPAE* 121:1-2: "In the year of hardship, whe[n[71] x *qû* of barley] went for a [shekel] of silver."

(e) *TSABR* 74:7: "In the year of hardship when two *qû* of barley (went for) a shekel of silver."

(f) *TSABR* 25:2-3: "In the year in which the TAR.PI army encamped against the city, (when) one *qû* of barley (went for) a shekel of silver."

Almost all of the Emar documents are undated; consequently, it is impossible to determine how frequently the formulae of "hardship" were used. However, the significant number of documents in which these formulae appear and the large number of scribes named as recording them suggest that this phenomenon was quite common.[72] The incorporation of the formula "in the year of hardship (and hostilities)" in the Emar documents, in which the recording of dates was not customary, strengthens the assumption that these formulae served a specific legal function, more so than in the Babylonian documents.

Several of the documents record the sale of children into slavery or their delivery for "adoption" by parents or other relatives (who probably served as guardians of the children[73]);[74] some of the transactions are the

70 *ki-i* 3 SÌLA ŠE.MEŠ *a-na* 1 GÍN KÙ.BABBAR *iz-za-az*. Compare the Panammuwa Inscription (*KAI* 215), line 6: וקם פרס בשקל, "a *prs* stood at a shekel"; cf. *b. Arakhin* 30a: "He sold it to him for a *mina* and he improved it so that its value now went for (lit. stood at) 200."

71 For this version, which corrects the reading of *RPAE,* see Tsukimoto, *ASJ* 10 (1988), 167. Tsukimoto restores here [3 *qû*] based on Documents (a)-(b) mentioned above; however, in light of the variations in the price of grain that we found in the Babylonian hardship formulae (see above, pp. 125-127) and also in documents (e)-(f) in this same section, it is better to refrain from completing the missing number.

72 Eight scribes connected with the hardship documents from Emar are known by name: Ea-mudammiq (*RPAE* 42), [Abi-]kapi (*ASJ* 12 [1990], 190 no. 7), Ehli-kusha (*RPAE* 149), Imlik-Dagan (*RPAE* 256), Ish-Dagan (*RPAE* 138, 139), Baal-gamil (*RPAE* 111), Dagali (*RPAE* 162; *ASJ* 12 [1990], 210 no. 16; cf. also *TSABR* 57), Tura-ahu (RE 31). In the other hardship documents – as in many of the legal documents from Emar – the names of the scribes are not recorded.

73 See *TSABR* 52: two people sell their niece into slavery for 12 shekels of silver.

74 *RPAE* 83, 216; *ASJ* 10 (1988), 166 Text E.

enslavement of adults in exchange for "sustenance" and the payment of debts which they were unable to meet "in the year of/at the time of hardship (and) hostilities."[75] The particularly difficult situation of women who remained without someone to support them at the time of hardship is reflected in the "adoption" documents, which specifically state that the woman "adopts" a man as a "son" and undertakes to make him the heir of her property so that he will support her "during the year(s) of famine."[76]

The hardship documents also record the sale of real-estate. In several such cases, the buyer is the king's son.[77] In the case of real-estate, it is more difficult (practically impossible) to infer the economic situation of the seller from the documents, since we cannot ascertain the relation between the price paid for the property and its true value.[78] Nevertheless, in cases where property was sold "in a year of hardship (and) hostilities" by women, we can assume that widowhood made their economic situation particularly difficult.[79]

75 *TSABR* 25: a man becomes enslaved to a person "who in the year in which an army surrounded the city and one *qû* of barley went for (lit., stood at) a shekel of silver," kept alive, sustained him by giving him 20 *parīsu* of barley, thereby saving him from starvation (the measure of a *parīsu* in the Hittite kingdom was 6 *sūtu*; see van den Hout, *RLA* 7.522–524). ASJ 13 (1991), 302, no. 37: a woman is enslaved to a man who "in the year of famine, when three *qû* of barley went for a shekel of silver," paid her debt of 25 silver shekels and kept her alive by enabling her to subsist "on bread and water during the year of famine." *RPAE* 121: a man is sold to one who paid his debt of 50 shekels of silver "during the year of hardship/famine [when x *qû* of barley] went for [a shekel] of silver" (cf. above, n. 71). Concerning *RPAE* 86, see below, p. 145, n. 85; compare also Tsukimoto, *ASJ* 10 (1988), 160–161, Text C, who reconstructs from the remnants of this fragmented document that the priests of Ereshkigal in Emar gave their temple to a man who "ke[pt them alive, sustained them] during the year of hardship."

76 *TSABR* 48, 74.

77 The purchase of real-estate by Iṣṣur-Dagan son of Pilsu-Dagan (king of Emar): *RPAE* 138, 139. On the purchase of real-estate by others, see *ASJ* 12 (1990), 207–209, Texts 15, 16; *RPAE* 20, 111, 149, 158, 162.

78 An exception to this rule is *TSABR* 65, from which it ensues that the price paid for a house that was sold in a year of hardship was no more than half of its true value; see below, p. 147, n. 89.

79 *RPAE* 20: a woman sells a house to her (step) son "in a year of hardship," after her sons had abandoned her; *TSABR* 65: a woman who was left with four children is forced to give up her house in return for payment of her debt. There is also an opposite example:

4. Parents Sell Their Children

As noted earlier, the end of *RPAE* 111 includes the formula "years of hostilities (and) hardship in which you sell the children whom you love, be it a daughter or a son."[80]

IV. THE LEGAL SIGNIFICANCE OF THE HARDSHIP FORMULAE

The ubiquity of hardship documents, their presence in Emar and in various Babylonian cities, and the fact that these were written by numerous scribes, indicates that documents of this type were widespread. Although there is no specific reference to situations of siege in the ancient Near Eastern law codices, the ubiquity of such documents makes it plausible that the hardship formula had a specific function. Yet, it is clear that the inclusion of these formulae was not obligatory, as follows from the existence of legal documents that were written during periods of siege (on the basis of the dates recorded therein) but do not contain the hardship formula.[81]

Why were the hardship formulae included in these documents?

The purchase of slaves and other property in periods of hardship was a form of investment that was both legal and profitable. The degree of profit to be expected from such a transaction is indicated by the following documents: K.153, a deed of sale from Babylon:[82] a woman said to

---

a woman buys a house from a man "in a year of hardship" for the price of 300 shekels of silver (*RPAE* 111).

80 See above, pp. 137–138.

81 Thus, for example, in two of the three documents from Nippur that record the purchase of children by the partners, Ardi-Gula and Iddina-Nergal (O.33, 47), the hardship formula appears, while a third document (O.54) makes no mention of any siege. Similarly, there is no allusion to siege in O.23, 26, which mention these two people and a third partner (see below). Cf. also Hunger, *BaM* 5 (1970), 236–237, no. 23 (Uruk); Ellis, *JCS* 36 (1984), 37 (NBC 4977); Weidner, *AfO* 16 (1952–53), 36–37, no. 1 (the two latter documents are from Babylon), in which the dates recorded indicate that they were written during a siege.

82 On this document, see further, Oppenheim, *ibid.*, 82–84; Dandamaev, *Slavery in Babylonia,* 175.

a man, "Keep me alive (i.e., sustain me)[83] and I will be your slave-girl (*bulliṭṭannima lū amatka anāku*)." He agreed and rationed her food. The document specifies that if her husband, son, brother or any other relative would offer to redeem her, saying, "I want to redeem (*apaṭṭar*) my sister," he [the relative] will give a replacement for her (lú*kūmišu inamdin*).[84] *RPAE* 83 discusses the case of a man who sold his daughter "[during the year of] hardship" in exchange for nine silver shekels. If he wished to redeem her, he would need to give her master two slaves (lit., "souls") in exchange. According to *TSABR* 52, two people sold their niece into slavery for 12

83 On the meaning of *bulluṭu* as "to provide food, sustain," see Oppenheim, *Iraq* 17 (1955), 71ff., esp. 82; CAD B *balāṭu* 7a. A similar sense is found for the verb חָיָה in 2 Kgs. 7:3-4: four lepers sit opposite the entrance to the besieged city of Samaria and say to one another, "Why should we sit here waiting for death? If we decide to go into the town, the famine is in the town and we shall die there; and if we just sit here still we die. Come, let us desert (נִפְּלָה) to the Aramean camp. If they sustain us we shall live, and if they put us to death, we shall but die (אם יְחַיֻּנוּ נחיה ואם ימיתֻנו ומתנו)." That is, sitting at the gate of the city or entering it would lead to certain death (of famine), whereas if they go over to the Aramean camp there would at least be a possibility that they would be fed (as is sometimes done with deserters, נופלים, lú*maqtūtu*, who are fed in order to get information from them; on the treatment of deserters, see Ephʿal, *EI* 24 [1993], 18-22). Thus, one should not interpret the word יְחַיֻּנוּ here, as do most of the commentators, simply in the sense of "leave us alive," for in such a case we would have to say that the same chance of being left alive awaited the lepers in this passage even if they do not go over to the Aramean camp. Compare also Ps. 33:18-19: "Truly, the eye of the Lord is on those who fear Him, who wait for His faithful care to save them from death, to sustain them in famine (להציל ממות נפשם ולחיותם ברעב)"; see further examples of this meaning of חָיָה in Lohfink, *Shaʿarei Talmon,* 111-119.

84 This siege document, from Year 19 or 20 of the reign of Shamash-shum-ikun, is the only extant Neo-Babylonian document in which a person sells himself into slavery (Dandamaev [*op. cit*., n. 82], 175). The mention of the woman's husband, son, brother and relative indicates that she was married and perhaps also a mother. The reference in this document to people closest to the woman may be explained on the assumption that her master knew of the existence of these relatives and of her separation from them due to the siege, and he wished to include them all in the section pertaining to the woman's right of redemption in the hope that he will get a high price (i.e., her full value) in return for freeing her.
It is interesting to note the difference in the fates of the two women mentioned in O.17 and K.153: the former managed to get a loan during the siege, conditional on returning the principle without interest "at the opening of the gate" (see below, pp. 147-151), whereas the latter was left with no option but to sell herself into slavery in exchange for food.

shekels of silver. If her father or her uncle wished to redeem her, they would need to provide two slaves in exchange. The people sold into slavery retained the right of redemption by their immediate relatives; however, the implementation of this privilege required payment of the full value of a slave, and at times even more (although the price which the owners paid at the time of purchase was much lower or that they received the slaves for free).[85] On the other hand, one should recall that the expenses incurred by the owners to sustain their slaves under siege conditions (when the price of food soared) was extremely high, whereas the exchange value of the slave after the siege ended was no greater than it was prior to it. Thus, the longer the siege continued, the anticipated profit to the buyers of slaves became smaller, and in the final analysis, they were even likely to lose money. So, the degree of profit the buyers expected to ultimately receive from their investment was in proportion to the degree of risk.

The economic nature of slave acquisition in periods of siege is also reflected in the existence of partnerships formed for this purpose. In "siege documents" from Nippur, from the third year of King Sin-shar-ishkun of Assyria, we find the following ratios between the number of purchasers and the price of children sold by their parents:

> One purchaser (Ninurta-uballit): 6 silver shekels (O.27, 30; recorded in Tammuz and Elul);
>
> Two purchasers (Ardi-Gula and Iddina-Nergal): 11 shekels (O.54; record of the month not extant); 12 shekels (O.33; the record of the month not extant); 15 shekels (O.47; recorded in Sivan);[86]

85 Yet it is worth noting the existence of documents written in times of hardship, in which the right of slave redemption is maintained without a double payment to their owners; see *TSABR* 25 (a slave in exchange for another slave); *ASJ* 13 (1991), 302 no. 37 (a female slave in exchange for a female slave); *RPAE* 86: a man is enslaved to a certain person who "kept him alive" (i.e., sustained him) during the "year of hardship," and paid back his debt of 2.5 silver shekels. If he wished to be freed from the house of his master, he would have to give him 10 silver shekels (the document does not state that the master should be given a slave in exchange for the freed slave).

86 Regarding the date of this document, see Brinkman and Kennedy, *JCS* 35 (1983), 58, "Year 3(+) not verified from cast." However, note that the Nippur documents from years 4–5 of Sin-shar-ishkun include no hardship formulae and documents from Uruk

> Three purchasers (the two mentioned above plus Ninurta-uballit): 22 shekels (O.24–26; recorded in Sivan); free (O.23; recorded in Sivan).

The relatively small amount invested by each of the partners (who occasionally received the slaves for free, as noted above) and the growing number of partners as the price of the children rises suggest that the risk to the investment was considered extremely high, for the life of the slaves was in danger and there was always the fear that they could die during the siege, an eventuality that would have rendered the entire investment lost.

It follows, therefore, that the inclusion of the hardship formula in the siege documents was intended to serve the buyers' interests, in order to prevent later demands by the sellers, that the children be returned to them at the end of the period of hardship for the original price paid, with a reasonable addition. The buyers therefore emphasized that the purchase had taken place under extraordinary conditions of high risk, hence they were entitled to realize their investment and to require the full price of the children paid (occasionally with additional profit) to secure their redemption.[87]

It seems, however, that this can only be a partial explanation of the purpose of the hardship formula, as these formulae also appear in other documents, which are not concerned with the sale of children and their conveyance to slavery, but with the sale of real-estate.[88] After the sale of real-estate in "times of hardship," there were no doubt sellers who were likely to argue that they had been exploited by sharp-witted buyers who

---

from years 6–7 of that same king include only the formula (*ina*) *edil bābi* (Brinkman and Kennedy, *ibid.*, 57–58).

87 On the legal and economic significance of the different forms of enslavement (self-enslavement due to starvation or due to unpaid debts) and of the liberation and redemption of slaves (whether as an investment intended to yield profit or with no intention of profit on the part of the owners), see Yaron, *RIDA* 3[e] serie, 6 (1959), 154–176.

88 The hardship documents from Emar, which include a particularly large number of documents dealing with real-estate, stress this problem even more. See, for example, Babylonian documents H.2; K.119, 140; and Emar documents *RPAE* 20, 138, 139, 149, 158, 162; *TSABR* 57, 65; *ASJ* 12 (1990), 189–190 no. 7; 207–208, no. 15; 208–209, no. 16.

took advantage of the severity of their financial situation; however, in this case, increased economic risk to the investment, the longer the siege continued, would not have been relevant justification as it was in the purchase of human beings.[89] Therefore, it cannot be argued that "hardship formulae" in documents recording the sale of real-estate were intended to prevent complaints of exploitation when the hardship ended. Moreover, several such documents indicate that the real-estate was purchased in times of hardship from "[the god] Ninurta and the elders of Emar,"[90] that is, from a public body that was not subject to personal suffering. In light of this finding, the question regarding the legal significance of the hardship formulae reemerges and remains as puzzling as ever.

V. "CHARITABLE DEEDS"

The documents written in Nippur during the final years of Assyrian rule over that city include three promissory notes,[91] in which the borrowers promised to return the principle[92] without interest[93] at "the opening of the gate."

Document O.17 (lines 18–19, 24–26) records a loan of a silver shekel to a woman and an additional quantity of silver (whose amount is not extant)

89 See *TSABR* 65; in this contract a woman and her four children sell their father's home "in a year of hardship" for 45 silver shekels. Should they wish to buy the house back, they would need to pay 90 shekels, thus suggesting that the price paid by the buyers was no more than half the real value of the house. This document is evidence of an overt intention of profiting from a war/hardship situation.

90 *RPAE* 139, 149; *ASJ* 12 (1990), 190 no. 7 (on this document, see above, p. 136, nn. 57–58).

91 The three documents discussed here make no explicit mention either of siege or of "closure of the gate." The determination of their political, economic and legal background is based on the dates mentioned in them and the phrase stating that the debts will be paid "at the opening of the gate." See Oppenheim, *Iraq* 17 (1955), 85. No mention is made in these documents of the term *ḫubuttūtu,* discussed below.

92 O.40: 3–4; 41:3–4: *ina patê bābi kaspa ana qaqqadīšu inamdin.*

93 O.17:26: *ša la ṣi[btu ina pat]ê bābi tanamdin.* The same formula evidently appears in lines 18–19 of the same document, which may be reconstructed as follows: *ša la ṣibtu [ina patê bā]bi* (K[À]) *ušallam.* On the structure of O.17, see the following note.

that was given to another person.[94] The other two documents (O.40 and 41), that are apparently later by several years (the date noted for one is 1st Marheshvan, year 6 of King Sin-shar-ishkun of Assyria, and the other notes the date 21st Marheshvan of the same year), record the names of the same borrower and lender. O.40 records that the borrower received one mina of silver, and according to O.41, he received an additional amount, one and a half mina of silver, twenty days later.

The obligation to return the money "at the opening of the gate" indicates that the borrowers had means and/or property which they could realize immediately, once the state of "closure of the gate" was ended, in order to

94 BR 8/7 63:24–26 is part of a *Sammeltext* which records, in five entries separated by lines, seven loans which were made by one lender, Rimut-Gula. Brinkman and Kennedy (*JCS* 35 [1983], 55) designate lines 18–26 in this document as O.17. In fact, this section contains three different loans: (a) lines 18–19: the borrower is ...*-uballiṭ* son of Bel-usati, who took a loan (whose amount is not extant) on condition of repaying it "without interest [at the opening of the g]ate" (see above, n. 93); (b) lines 19–23: a loan that was given in the month of Tammuz, year 2 of Sin-shar-ishkun, to a woman and her daughter whose names appear in line 21. Lines 22–23 list the names of the witnesses and the scribe, as well as the date, 10th of Shebat, year 2 of Sin-shar-ishkun (the fact that a loan given in Tammuz was only documented in Shebat, eight months later, calls for explanation). This loan was given at 20% interest, as noted in the heading of the *Sammeltext*; (c) lines 24–26 record a loan of one shekel of silver given to Ahat-eresh, daughter of Apla, on condition that it will be returned without interest "[at the openi]ng of the gate." Note that loan (b) includes the names of the witnesses and the scribe as well as the date, as is customary in documents of economic transactions, while these important details are not recorded in the details of loans (a) and (c). Only partial explanation may be offered for this phenomenon: since both loans were exempt from interest, the date was of no real significance; this, of course, does not explain the absence of the names of the witnesses and the scribe. Another possible explanation is that the scribe and the witnesses listed served to confirm all three loans on the same date. However, this possibility leaves open the question why the names of the scribe and witnesses and the date appear at the end of document (b) and not at the end of the final document (c).
The dates of the other loans in the document are year 3 of Ashur-etel-ilani and year 2 of Sin-shar-ishkun, kings of Assyria. There is no notation of the original place where the documents copied here were written. On the considerations supporting their origin as Nippur, see Borger, *WZKM* 55 (1959), 67–68; *idem*., *JCS* 19 (1965), 65–66; von Soden, *ZA* 58 (1967), 245–246. The quantities of silver given in two of the other loans recorded in this document are in the amount of one mina to each borrower; the quantities of the four remaining loans are not extant.

repay the debt. We can speculate that the woman who borrowed a shekel of silver needed it for survival, yet she managed to obtain a loan rather than sell herself into slavery due to the gravity of her situation.[95] By contrast, it appears that the large amount of silver (150 shekels) received by the borrower in O.40 and O.41 was not required for daily living expenses, but rather was used to finance economic transactions[96] or for exceptional survival needs. It appears that there was a certain family connection between the borrowers and the lenders, which rendered the loan "a charitable deed," rather than a regular financial transaction based on the loaner's profit margin. Several points indicate the familial nature of the transactions between Ninurta-gamil son of Apla (the borrower) and Ninurta-ushallim son of Nabu-usipi (the lender) in O.40–41: the borrower was also the scribe who wrote O.40[97] and the only witnesses mentioned in this document are two sisters (especially in light of the fact that women only rarely served as witnesses in Neo-Babylonian legal documents of this period).[98]

In this context, it is worth mentioning additional loan documents written in times of hardship, which record interest-free loans according to some scholars:

K.133 was written in Babylon on 9 Tebeth, year 19 of Sin-shar-ishkun.[99] In this document, the borrowers, a man and his wife, undertake to repay the loan in full (*ina qaqqadīšu*),[100] that is, the principle, in the amount of 50 silver shekels, "on the day when (access to) the country is opened (again)" (*ina ūme pān māti ittaptû*). The loan is described in this document as *ḫubuttūtu*.

T.5.6 is a loan document from Uruk, dated 16th Ab of the 5th year of Nabopolassar of Babylon. The document ends (line 16) with the formula

95 Unlike the case discussed in K.153 (see above, pp. 143–144).

96 On transactions of this type, see above, pp. 143–146.

97 Note that O.41 was written by a different scribe.

98 See Dandamaev, *Slavery in Babylonia,* 48.

99 At the end of the document, ll. 18–21, appears the following hardship formula: "In those days the city was besieged. Famine and hardship (were) in the country. People died for want of food."

100 Literally: "on his head." For this meaning, compare Lev. 5:24: "He shall repay the principal amount (בראשו, lit., its head) and shall add a fifth to it."

ina edil bābi ("at the closing of the gate"). This document states that the three minas of silver which were given to the borrower are considered as *ḫubuttūtu* from the 16th of Ab (until) *adi balāṭ*, and that if the loan is not returned *ina balāṭ,* it will incur interest in the amount of 6 shekels per mina (10%). Oppenheim initially proposed that the expression *ina balāṭ* (lines 3, 5) refers to "reviving/saving" the city, that is, to the cessation of the siege during which the transaction took place.[101] He later retracted this interpretation, as it became clear that *ina balāṭ* is simply the equivalent of the Hebrew כעת חיה (i.e., next year at this time).[102]

The pertinence of these documents to the present topic depends on the interpretation of the term *ḫubuttūtu,* which is itself debatable. Since the term appears in some of the documents alongside explicit designation that only the principle (*qaqqadu*) is to be returned, some scholars suggest that the meaning of the term *ḫubuttūtu* is an interest-free loan.[103] Others argue that *ḫubuttutu* means "the amount of debt including interest" (according to this view, the term *ina/ana qaqqadi* does not refer to the principle alone, as is usually thought, but rather to "the amount of debt in full, without the addition of compounded interest").[104] According to this approach, the payment of "the debt in full," *ina qaqqadīšu,* already includes calculation of interest, thereby satisfying the lender's interests in making a profit; accordingly, the transactions in K.133 and T.5.6 are not "charitable deeds."

The formulation of T.5.6 is open to both interpretations: according to the first approach,[105] this is an interest-free loan for a period of one year,

101 *Iraq* 17 (1955), 78.

102 See CAD B 51, *balāṭu* 3.

103 See the definition of *ḫubuttūtu* in *AHw* 352 (anfangs zinloses) Darlehen; CAD B (1965), 51 s.v. *balāṭu* 3: loan without interest. See also CAD Ḫ (1956), 221 s.v. *ḫubuttūtu.* Although the discussion mentions Bilgiç's differing view (see next note), the dictionary does not tend to accept it.

104 See Bilgiç, *Ankara Universitesi Dil ve-Tarih-Coğrafya Fakültesi Dergisi* 5 (1947), 451-452; Shiff, *JCS* 40 (1988), 187-194. For the assumption of etymological parallelism between the terms *ḫubuttūtu* and נשך, see Lipiński, *OLP* 10 (1979), esp. 138, n. 30, and references therein.

105 See above, n. 102.

given in a time of siege out of the lender's good will. The 10% interest is only calculated after the first year of the loan. According to the second approach,[106] the 3 minas recorded in the document as the borrower's debt already include the regular interest (whose amount is not mentioned), while the 10% interest is a penalty to be added if the debt is not repaid within a year. The formulation of K.133, however, raises significant doubts regarding the second interpretation, since the interest cannot be calculated in advance and added to the principle when the end of the siege ("the day when [access to] the country will be opened") is unknown.

106 See above, n. 103.

CHAPTER FIVE

SOCIAL ASPECTS

I. BETWEEN MAN AND GOD

Various religious ceremonies and activities were probably conducted in a besieged city. Unfortunately, there is very little extant data on this subject, most of it from the western part of the Fertile Crescent and related cultures. According to our sources, appeals to the god(s) were generally initiated by the king, who personally participated in such ceremonies.

Appeals to the god(s) can be divided into two spheres: appeals within the ceremonial sphere of prayer and request for deliverance, and appeals within the mantic sphere of divination. The technical nature of appeal to the god(s) suggests that the activities required specialized experts: priests in the sphere of prayer and various kinds of prognosticators in the sphere of divination. In the prophetic sections of biblical literature, in which the prophets play a central role in the life of the people, these functions were merged and the prophet performed both tasks: praying for deliverance on the one hand and delivering responses to appeals to God on the other.

1. Prayer

The prophetic story of the deliverance of Jerusalem at the time of Sennacherib relates that, after listening to the emissaries of the king of Assyria, King Hezekiah of Judah went to the Temple. According to the biblical reports (2 Kgs. 19:1 and Isa. 37:1), Hezekiah came to the Temple in sackcloth, but what he did there is not reported. According to 2 Kgs. 19:15–19 and Isa. 37:15–20, Hezekiah prayed to God "enthroned upon the Cherubim" (possibly in the inner sanctuary, in which the Ark of the Covenant was kept) and appealed to Him to save Jerusalem from Sennacherib.[1] Similarly, the inscription of Zakkur, king of Hamath and Luʿash, after giving details of the military coalition which had placed Hadrach under siege, states: "I lifted my hands to Baalshamayn, and Baalshamayn answered me, and Baalshamayn [spoke] to me by seers and messengers (ביד חזין וביד עדדן)."[2] A document from Emar states that when the Hurrian army laid siege to Emar, the king, Pilsu-Dagan, son of Baal-kabar, raised his eyes to Baal (*ᵈX*),[3] who gave a favorable sign (possibly in the flight of birds),[4] and went on to defeat the soldiers (of the Hurrian king) who were already within the city and on its wall, and saved Emar.[5]

2. Child Sacrifice

According to 2 Kgs. 3, at the end of a passage concerning a joint campaign of Israel and Judah in the land of Moab and a siege they placed on the Moabite capital, Kir-haresheth, it is said: "Seeing that the battle was going against him, Mesha the king of Moab led an attempt of seven hundred swordsmen to break a way through to the king of Edom. So he took his firstborn son, who was to succeed him as king, and offered him up on the

1 It is customary to draw a distinction between two sources dealing with the same subject in the prophetic story concerning Sennacherib's war against Judah (the first in 2 Kgs. 18:17 – 19, 9a, 36 // Isa. 36:2 – 37, 9a, 37, the second in 2 Kgs. 19:9b-35 // Isa. 37:9b-36). The first source was composed close to the time of the events; the second seems to have been composed two or three generations after the former and includes a prayer ascribed to Hezekiah that seems suited to the Deuteronomistic school. In both sources, the prophet Isaiah is the center of the story; cf. Cogan and Tadmor, *II Kings*, 240-244.

2 See *KAI* 202 A 11ff.

3 On the king raising his eyes to Baal, compare below, *CAT* 1.119.26–33 (translation of line 27': "lift your eyes to Baal" and line 34': "[And Ba]al will hear [your] prayer."

4 Document *ASJ* 12 (1990), 190, 7:29-37 speaks of the salvation of Emar from siege following an extispicy by a diviner (ˡᵘ́*bārû*); the document discussed here (*RPAE* 42), possibly related to the same event, mentions a different technique of divination, possibly based on the flight of birds – if the reading *ù* ᵈX *eg-re-ti* MUŠEN GI *ša* ŠÀ-*šu i-di-na-šu* is correct. For the response of the god given by means of several techniques, cf., the Zakkur Inscription, *KAI* 202 A 12.

5 See *RPAE* 42:8-16. For textual corrections and discussion, see above, pp. 135-137, nn. 59-61.

wall as a burnt offering. A great wrath came upon Israel, so they withdrew from him and went back to [their own] land" (vv. 26–27). The nature of this wrath is not explained in the text, and it is difficult to determine the significance of Mesha's act. One possibility is that the sacrifice was intended to arouse the wrath of Chemosh, the god of Moab, thereby forcing the armies of Israel to retreat; another possibility is that the text refers to the wrath of YHWH, God of Israel, which rose due to the sacrifice offered by Mesha to his own god; according to a third option, this verse simply refers to the anger of the Moabite combatants following their king's sacrifice, which fuelled their attack and allowed them to repel the besieging army. In any event, it is clear that the author saw the wrath in question as the cause for the Israelite withdrawal from Moab; thus, from Mesha's perspective, his sacrifice achieved the desired result.

The custom of child sacrifice, particularly in times of adversity, is mentioned in various sources pertaining to the Phoenician culture. Thus, for example, Philo of Byblos states that in times of adversity, it was customary among ancient peoples for rulers or leaders to sacrifice their most beloved children to the vengeful gods, and that these children were put to death in a mysterious ceremony.[6] Diodorus relates that after the defeat of Carthage to Agathocles, the tyrant of Syracuse, which left the city under siege (in the year 310 BC), the Carthaginians examined their behavior to determine the cause of their suffering. Finally, they concluded that the god Chronos brought this evil upon them because they had neglected the time-honored custom of offering their finest sons as sacrifice in his honor, and instead sacrificed children that were secretly purchased. In their haste to correct this lapse, the Carthaginians chose 200 of the most outstanding children and sacrificed them in public, and at least another 300 people (suspected of offering up children other than their own) sacrificed themselves voluntarily.[7]

6 In Eusebius, *Praep. Evang.*, 1.10.44 = 4.16.11. If the work of Philo of Byblos is indeed based on Sanchuniathon's Phoenician History (for the scholarly debate on this question, see Attridge and Oden, *Philo of Byblos: The Phoenician History*, 3–9, and 93 n. 148), then the information presented here is of particular importance owing to the antiquity of the period discussed.

7 Diodorus 20.14.4–6. On other acts of the Carthaginians during that siege, see below, n. 11.

Similarly, we are told by Curtius Rufus that during Alexander's siege of Tyre, there were those in Tyre who suggested reviving the custom of sacrificing a youth to Saturn, a custom that had been abandoned many years earlier.[8]

Accounts of Carthage are remote in time and place from the framework in this book, and their relation to the eastern Mediterranean in the period discussed is indirect and requires confirmation. There are, however, additional sources, relating to Phoenicia and Canaan during the 13th and 12th centuries BC, which should be taken into consideration in the discussion of child sacrifice during siege.

(I) A Ugaritic text describes a ceremony that should be conducted in a city when enemies appear at its gates or during a battle over the city:

k gr ʿz ṯǵrkm qrd
ḥmytkm ʿnkm l bʿ*l tšun*
*y b*ʿl*m* [a]*l tdy ʿz ṯǵrn*
y qrd [l]ḥ*mytny ibr y*
*bʿl nš*q*dš mḏr bʿl*
nmlu b/d*kr b*ʿl *nš*[q]dš
ḥtp bʿl nmlu ʿšrt bʿl n[ʿ]
šr qdš bʿl nʿl ntbt bt [bʿl]
ntlk w šmʿ [b]ʿ*l l ṣlt*[km]
ydy ʿz l ṯǵrkm [qrd]
*l ḥmytk*m[9]

8 Curtius Rufus 4.3.23. Cf. Plutarch, *Moralia,* 175a, who relates how the tyrant Gelon, after defeating the Carthaginians at Himera (480 BC), forced them to sign a peace treaty in which they pledged not to offer up their children to Chronos. For a collection of (translated) passages on the Carthaginian custom of child sacrifice from the writings of the ancient historians and the Church Fathers, see Day, *Molech: A God of Human Sacrifice in the Old Testament,* 79-81.

9 *CAT* 1.119.26-36; cf. Pardee, *Ritual and Cult at Ugarit*, 52-53. For discussion of this text, cf. Margalit, *Proceedings of the Seventh World Congress of Jewish Studies: Bible and Ancient Near East, 63-83;* Avishur, *Studies in Hebrew and Ugaritic Psalms*, 253-276; Saracino, *ZAW* 95 (1983), 263-269.

Translation (with some modifications):

> (26') When a strong foe attacks your gate,
> a warrior (27') your walls,
> You shall lift your eyes to Baal and say:
> (28') O Baal, if you drive the strong one from our gate,
> (29') the warrior from our walls,
> A bull, (30') O Baal, we shall sanctify,
> a vow, O Baal, (31') we shall fulfill;
> a firstborn (or: a ram; see below), O Baal, we shall sanctify,
> (32') a *ḥtp*-offering, O Baal, we shall fulfill,
> a feast, O Baal,[10] we shall (33') offer;
> To the sanctuary, O Baal, we shall ascend,
> that path, O Baal, (34') we shall take.
> And Ba[al will h]ear [your] prayer:
> (35') He will drive the strong foe from your gate,
> [the warrior] (36') from yo[ur] walls.

Lines 28' to 33' contain a declaration of the obligation to fulfill a vow to Baal. Its exact nature depends, however, on the reading of the word *[b/d]kr* in line 31', which is debated. If the word *[b]kr* (firstborn) in line 31' is correctly reconstructed, then the text can be understood as reference to an obligation to sanctify (possibly meaning "to sacrifice") the first born son to Baal (compare Micah 6:7: "Would the Lord be pleased with thousands of rams, with myriads of streams of oil? Shall I give my firstborn for my transgression, the fruit of my body for my sins?").[11] It should nevertheless

10 On the suggestion that *ʿšrt bʿl* here may mean "a tithe to Baal" rather than "a banquet offering to Baal," see Greenfield, *ʿAl Kanfei Yonah*, 764, n. 22.

11 Compare Diodorus 20.14.1-2, who reports that from the beginning of settlement in Carthage, the Cartaginians would send a tithe of all public income to the god Heracles (Melqart?) in Tyre, their city of origin. In the course of time, as they became wealthier, their contributions to the god became much smaller than a tithe. When Agathocles laid siege to Carthage, its inhabitants believed that the disaster was a result of Heracles' anger over their negligence; therefore they sent a large sum of money and numerous precious gifts to Tyre. For Diodorus' account on child sacrifice in connection with the siege, see above n. 7.

Fig. 11: The Conquest of Ashkelon. Fighting in the gate area.
Relief from Karnak, from the reign of Merneptah
(relief was previously attributed to the period of Rameses II)
From Wreszinski, *Atlas zur altägyptische Kultursgeschichte*, II, Taf. 58

be noted that, from a palaeographic viewpoint, there is a certain difficulty in this reconstruction, and that the correct reading may in fact be *[d]kr* (male animal, a ram);[12] if so, this text does not deal with human sacrifice at all.

12 See Xella, *RSF* 6 (1978), 127-136 and the bibliography mentioned above, n. 9.

(II) A regular scene depicting battles against Canaanite cities appears in Egyptian reliefs from the reigns of four kings of the Nineteenth Dynasty (13th and early 12th century BC). The scene shows the Pharaoh and his army attacking a city while the defenders stand on the walls (the inner walls, if the outer walls are under attack), their hands raised in prayer,[13] one of them holding a burning censer, and a child (sometimes two) being lowered from the wall. Spalinger interprets this as depicting a ceremony of child sacrifice on the walls, accompanied by prayer, similar to the acts of Mesha and the other passages cited above.[14] In light of all these sources, it is difficult to accept Keel's interpretation, that incense is offered by the besieged in recognition of Pharaoh's theophany and as a declaration of loyalty to him, and that the children lowered from the wall are the children of the city's leaders, given to Pharaoh (possibly as hostages) to indicate their parents' loyalty.[15]

In sum, it must be reiterated that the evidence concerning child sacrifice in times of adversity clearly refers only to the Palestinian and Phoenician-Punic region. Evaluating the shock value of child sacrifice at times of adversity depends, to a large extent, on the information available regarding the frequency or rarity of the custom in the region at times of non-adversity, particularly in the worship of Molech. In the debate as to whether human sacrifice was indeed customary, as the evidence seems to suggest at face value, some scholars argue that the biblical passages detailing such activities (fathers passing their sons and daughters through fire, setting them on fire or burning them) should not to be understood literally. Rather, they argue, the biblical passages refer to a symbolic act that did not harm the children.[16] Excavations of the Topheth in Carthage, and in other sites in North Africa, Sicily and Sardinia, revealed the remains of thousands

13 Compare the Zakkur Inscription, line 11: "I lifted up my hands to Baalsha[my]n and Baalshamayn answered me."

14 Spalinger, *JSSEA* 8 (1977), 47–60.

15 Keel, *VT* 25 (1975), 413–469.

16 See Weinfeld, *UF* 4 (1972), 133–154. For a survey of views regarding the significance of Molech worship in Judah, and supporting the view that children of both sexes were indeed sacrificed to Molech in Judah (in the absence of adversity), see Day (above, p. 155, n. 8); cf. below, n. 18.

of burned children, alongside memorial stones mentioning their sacrifice to the goddess Tanit;[17] in light of these discoveries, there would seem to be little room for doubt as to the existence of the practice.[18] The evidence cited, in conjunction with the biblical story of Jephthah's daughter and the oath he made (Judg. 11:30–31, 34–39), indicate beyond a doubt that child sacrifice was customary in times of adversity, and possibly even on a regular basis.

3. Appeal to the God(s) during Siege

One of the characteristic acts of the king of a besieged city is his appeal to the god through prayer and divination. The historiographic sources sometimes emphasize that such an address – followed by deliverance – was made at a fateful moment, just before the fall of the city: Zakkur, king of Hamath and Luᶜash, lifted up his hands to Baalshamayn when his enemies had completed the preparatory stages for the decisive attack on Hadrach;[19] Mesha, king of Moab, offered his son as a burnt-offering on the wall of Kir-haresheth once he became aware that "the battle was going against him," and after the failure of a sortie he had made with "seven hundred swordsmen to break a way through to the king of Edom" (2 Kgs. 3:20–27); the Hurrian army, which was already on the wall of Emar and within the

17 It should be noted that according to the memorial stones, the children were offered to Tanit, while the literary testimonies speak of sacrifice to Chronos and Saturn (both names refer to the same god).

18 Excavations at Carthage suggest that child sacrifice was customary there over a long period of time (ca. 725–146 BC); see Stager, *Qadmoniot* 17 (1984), 39–49. See Mosca, *Child Sacrifice in Canaanite and Israelite Religion* for a detailed discussion of the classical sources, and particularly of the different terms appearing on the Punic memorial stones and in the Bible, leading to the conclusion that child sacrifice was customary and institutionalized in the Phoenician-Punic world, even in the absence of adversity, and that the Bible provides evidence suggesting the existence of this practice in Israel as well. For discussion of the Punic memorial stones connected with child sacrifice (as well as for a list of Classical and early Christian sources relating to this practice and an enumeration of the sites in which these were discovered together with urns containing half-burnt bones of children), see Brown, *Late Carthaginian Child Sacrifice and Sacrificial Monuments in their Mediterranean Context*.

19 See above, p, 81.

city, was defeated when King Pilsu-Dagan of Emar lifted up his eyes to Baal.[20]

The king's appeal was made through experts in divination (prophets, חזין and עדדן, *bārê*, etc.).[21] Sometimes the king received an unfavorable response; such was the case of King Zedekiah of Judah, who sent delegations to the prophet Jeremiah during the siege of Jerusalem, asking him "please inquire of the Lord on our behalf, for King Nebuchadrezzar of Babylon is attacking us; perhaps the Lord will act for our sake in accordance with all his wonders, so that [Nebuchadrezzar] will withdraw from us" (Jer. 21:1-2), and later: "Please pray on our behalf to the Lord our God" (*ibid.* 37:3).[22] The prophet's response that "the army of Pharaoh which set out to help you" will not save Judah, that God has determined the destiny of Jerusalem to be consumed in fire by the Chaldeans and that the only survivors would be "whoever leaves and goes over to the Chaldeans who are besieging you shall live, he shall at least gain his life" (Jer. 21:9; compare 37:7-10).[23] However, most cases mentioned in our sources include a favorable response from the god, followed by salvation. As noted above, the inscription of Zakkur, king of Hamath and Luᶜash, relates that on the decisive moment: "I lifted up my hands to Baalshamayn, and Baalshamayn answered me, and Baalshamayn [spoke] to me by seers and messengers (ביד חזין וביד עדדן),[24] and Baalshamayn [said to me], Fear not, because it was I who made you

20 See above, p. 153, n. 5.

21 For other ways in which the king appealed to the god in times of national adversity, compare Saul's appeal to the prophet Samuel through the witch of En-dor on the eve of the decisive battle with the Philistines (having failed to obtain an answer for his appeals to God "either by dreams, or by *Urim*, or by prophets"), a battle in which Israel was in fact defeated and Saul found his death (1 Sam. 28:4-19).

22 From the prophet's response: "Thus said the Lord God of Israel: Thus shall you say to the King of Judah *who sent you to me to inquire of me...*" (Jer. 37:7), it follows that Zedekiah did not send a delegation merely to "pray on behalf of...," but in order to ascertain that there was "a word from the Lord" (compare Jer. 37:17); cf. also "...pray for us to the Lord your God... Let the Lord your God tell us where we should go and what we should do" (Jer. 42:2-3).

23 Jeremiah's prophecy was included in the biblical canon after it became clear that it was fulfilled, thus indicating that it was a true prophecy (compare Deut. 18:18-22).

24 On the title עדדן as designating a particular type of diviner, see Uffenheimer, *Early Prophecy in Israel*, 32-39.

king, [and I shall stand] with you, and I shall deliver you from all [these kings who] have forced a siege upon you" (*KAI* 202 A 11–15). Similarly, after receiving Rabshakeh's ultimatum, King Hezekiah of Judah sent a delegation to the prophet Isaiah, among them the official in charge of the palace, the scribe and the senior priests, "covered in sackcloth… They said to him: 'Thus says Hezekiah: this day is a day of distress, of chastisement and of disgrace. The babes have reached the birthstool, but there is no strength to bring them forth (i.e., the situation is desperate and we are at a loss). Perhaps the Lord your God will take note of all the words of the Rabshakeh, whom his master the king of Assyria has sent to blaspheme the living God, and will mete out judgment for the words which the Lord your God has heard – if you will offer up prayer for the surviving remnant'" (2 Kgs. 19:2–4 // Isa. 37:2–4). In response, Isaiah prophesies in God's name: "Do not be frightened by the words you have heard, ..." (*ibid.*, vv. 6–7).

The combination of an appeal to a god during a siege and the receiving of a (favorable) response from the diviner appears in two documents from Emar. According to one of these, when the Hurrian army laid siege to Emar, its king, Pilsu-Dagan son of Baal-kabar, raised his eyes to Baal (^{d}X),[25] who gave him a favorable omen (possibly in the flight of birds). The king then defeated the soldiers of the king of the Hurrians who were within the city and on the wall, and saved the city.[26] Additional details regarding this event can probably be found in the second document, a deed of gift stating that when the Hurrian army laid siege to Emar, the divination of Mashruhe, the diviner (^{lú}*bārû*) of the king and of the city, was fulfilled, and therefore King Pilsu-Dagan granted him a field.[27]

25 In addition to the simple visual aspect, the idiom, ...נשא עיניים אל ("to raise one's eyes to…") does not necessarily designate prayer, but can also indicate an act of looking up with the expectation of forgiveness or clemency. Compare: "To You, enthroned in heaven, I turn my eyes (נשאתי את עיני). As the eyes of slaves follow their master's hand, as the eyes of a slave-girl follow the hand of her mistress, so are our eyes toward the Lord our God, awaiting His favor" (Ps 123:1–2). The expression *īna našû,* in addition to simply looking, means "to desire, to want" (see CAD N/II 104, *našû* A 6, sub *īnu*). As noted earlier, in the inscription from Ugarit and in the document from Emar cited here, this idiom means "to pray."

26 See *RPAE* 42 8–16, and see further above, p. 153, n. 5.

27 Tsukimoto, *ASJ* 12 (1990), 190:29–33, and the discussion therein, pp. 191–192; cf.

II. PUBLIC LIFE

1. Maintaining Morale and the Handling of Hostile Elements

Unlike pitched battle, which requires exhilaration and willingness to engage in a one-time, brief effort, a siege requires continued determination and discipline over an extended period of time from both the besieged and the besiegers (and, in the case of a city, maintaining public order as well). A city whose defenders – and to a large extent its non-combatant population as well – are not united, cannot hold its own even if its fortifications are strong. The same holds true for the besieging army: if the warriors' morale flags, the army will soon find itself forced to forego its goal. The maintenance of discipline and morale depends on a complex mix of psychological, social and economic factors. Our knowledge of these factors in every period and locale is partial and fragmentary. This is particularly true regarding the ancient Near East, since nearly all extant sources on these matters, other than the Bible and the writings of some ancient Greek authors, are associated with the royal establishment and do not necessarily reflect the mood of the general population. The information available about public life in the ancient Near East during siege is, therefore, sparse and fragmentary, and its interpretation often requires conjecture.

The reality of a besieged city is characterized by extreme situations, demographically, socially and economically. As a rule, many foreigners lived in the city, often as many as the local inhabitants, but were not included among the local population. These included inhabitants from the nearby environs, people from unprotected settlements and semi-nomads who fled to the city in search of protection within its walls,[28] military personnel whose troops retreated into the city and others.[29] The presence of foreigners in the

---

above, p. 153, n. 5. As stated above, almost all of the writings referring to an appeal to the god during a siege appear in literary sources that describe deliverance coming at the decisive moment. The gift document mentioned in this note is the only non-literary source, from which one may learn that appeals to the god at such times indeed took place, through the mediation of a diviner.

28 Such as the Rechabites (Jer. 35:10–11), the prophet Jeremiah and his cousin Hanamel, who owned land in Anathoth (Jer. 32:6 ff.).

29 There are various examples for this: the Canaanite kings and their people, who gathered

city required their prompt assimilation into everyday urban life in terms of provision of food, water and shelter (which they lacked entirely), as well as constant supervision of their behavior. According to the guidelines of Aeneas the Tactician, any aliens found within a besieged city should be disarmed (10.9); they should not be housed without the permission of the authorities, who must have a list of all alien lodgings (10.9); they should be prevented from wandering about at night (10.10) or from observing the drill of mustering combatants within the city (10.13). Foreign mercenaries are to be under particularly close supervision – since they are trained and may be organized into military units, they represent a potential threat to the inhabitants of the city. It is therefore important to ensure that the citizens (i.e., inhabitants of the city) who invited these mercenaries exceed them in number and in power, otherwise they and the state will be at the mercy of the mercenaries (12.2, 4). Even allied forces should be divided and not allowed to camp together within city limits (12.1). Mercenaries must be warned that any harm they may cause to the military activity of the city, including damage to morale, will be considered a crime punishable by death (10.19).

Danger was also to be expected from certain sectors of the local population – groups and individuals who oppose the leadership of the city and its political stance.[30] It is also possible that members of the economically poorer and more vulnerable classes constituted a dangerous element, since for them a siege could lead to extreme hunger, the loss of property to the point of enslavement and the selling of their own children. The potential

---

in Megiddo following a defeat in battle against the army of Thutmosis III which took place in the vicinity (Wilson, *ANET*, 236–238); the Arabs who stayed in Babylon when they came to the aid of Shamash-shum-ukin (Streck, *Asb*., 68 viii 35–41); the *urbi* and the *tillāti* warriors brought to Jerusalem by Hezekiah to strengthen his troops during Sennacherib's campaign against Judah in 701 BC (on the etymology of these terms and the identity of the groups referred to, see Tadmor, *Beer-sheva* 3 [1988], 171–178); compare also Hazael's retreat into (besieged) Damascus, after he was defeated by Shalmaneser III in the region of Mt. Senir in 841 BC (RIMA 3 48).

30 Compare, for example, the statement of Rib-Haddi, ruler of Byblos, that half the inhabitants of his city supported the sons of Abdi-Ashirta, and the other half were loyal to Pharaoh (*EA* 138:71–73).

danger embodied in these groups lies in the possibility of conquest through betrayal (*sartu*), rebellion (*bārtu, sīḫu*) and upheaval (*nabalkattu, gabaraḫḫu*), as recorded in the queries to the Mesopotamian god(s),[31] and in stories in Greek and biblical sources concerning collaboration with the enemy,[32] desertion[33] and causing discouragement (lit, "weakening the hands..."). Consequently, emphasis is placed on the need to watch these opposition groups, to neutralize them and to prevent them from acting as a "fifth column."[34]

Fears of the damage that may be caused to the besieged city and its ability to withstand required the isolation of agitators and dissenters to

31 See above, p. 22.

32 Thus, for example, the tradition concerning the conquest of Canaan by the Israelites tells of Rahab the harlot who sheltered the spies who were sent to investigate the land and Jericho (Josh. 2); a similar tradition tells of the House of Joseph, who captured Bethel after one of its inhabitants showed them the way into the city in exchange for sparing his life and the lives of his family (Judg. 1:23–26).
A striking example of a leader's treachery is the story of Tennes, king of Sidon, who rebelled against Artaxerxes III, king of Persia, but when the Persian army came to Phoenicia, he secretly negotiated with them and handed the leaders of the rebellious city over to them (Diodorus 16. 43.1–4; 45.1–4).

33 According to Jer. 38:19, "the Judeans who have defected to the Chaldeans" (cf. 2 Kgs. 25:11; Jer. 39:9; 52:15) are mentioned as a group of whom Zedekiah was afraid and therefore did not turn himself over to the officers of the King of Babylonia, lest "they (the Chaldeans) might hand me over to them to rebuke me." The phrasing implies that Zedekiah considered this group as traitors who had gone over to the enemy not only due to the hardships of war, but out of animosity toward him personally. According to Jer. 52:29, Nebuchadnezzar, in the 18th year of his reign, had taken captive "from Jerusalem eight hundred and thirty-two persons." But the city itself did not fall to the Chaldean forces until the fourth month (Tammuz) of the 19th year of Nebuchadnezzar's reign (cf. 2 Kgs. 25:8; Jer. 39:2; 52:6–7, 12). One may therefore identify the exiles "from Jerusalem" (as opposed to other groups of exiles, referred to as "the Judeans" in Jer. 52:28, 30) with "the Judeans who have defected to the Chaldeans" long before the fall of the city, and whom Zedekiah feared. Thus, it would seem that Jer. 52:15, 29 deals with different groups of deserters.
Another group of deserters is mentioned in the story of the four lepers at the gate of besieged Samaria (2 Kgs. 7:3ff.); on these, see above, p. 144, n. 83.

34 For instructions intended to increase supervision of potential conspirators and rebels and to reduce their potential to cause damage, see: Aeneas the Tactician 1.9; 10.4–5, 7, 15; cf. also 22.6, 14; 28.2.

prevent contact between them and the other inhabitants. The laws in the ancient Near East make no mention of imprisonment as punishment for criminal or religious transgressions, but it is mentioned in the historical literature as a method used by the authorities to isolate elements considered dangerous to the public.

An interesting example of the steps taken against deserters and those that damage the public morale is found in the story of Jeremiah's imprisonment during Nebuchadnezzar's siege of Jerusalem in the reign of Zedekiah (588–586 BC). Jeremiah was imprisoned in the prison compound (חצר המטרה), where he remained until the fall of the city. The book of Jeremiah provides several versions of the circumstances of his imprisonment, outlined below.

According to one version (Jer. 37:5–21), Jeremiah was caught by an inspector/an officer of the guard (בעל פְּקִדֻת; *bēl piqitti*) at the Benjamin Gate while attempting to leave the city during a lull in the siege, "when the army of the Chaldeans raised the siege of Jerusalem on account of the army of Pharaoh." The dispute between the two revolved around the prophet's *intention* in leaving Jerusalem – whether, as he claimed, he only intended "to escape from there among the people" (לַחֲלִק משם בתוך העם), that is, to escape from the city and the suffering induced by the siege, or, as Irijah, the officer of the guard accused, he intended to go over to the Chaldeans (אל הכשדים אתה נֹפל), that is, to desert and give himself over to the enemy (37:12–21).[35] The officials before whom Jeremiah was brought beat him and threw him into "the pit and into the cell," where he stayed for many days. One may suppose that the officials settled for detaining Jeremiah rather than punishing him in full for such a serious offence, since desertion could not be proven as long as it was not executed in practice. The prophet's plea to Zedekiah, "Now, please hear me, O lord king, and grant my plea: Do not send me back to the house of the scribe Jonathan or I shall die there" (37:20), implies that this incarceration was under harsh conditions. Following his plea, Jeremiah was removed from this strict place of

35 For details on the background to this incident, including the meaning of the verbs לחלק < *ḫalāqu*, and נפל אל/על (and its Akkadian parallel *maqātu ana*), see Ephʿal, *EI* 24 (1994), 18–22.

imprisonment and transferred, at the king's order, to the prison compound, where he was given a daily ration: "a loaf of bread from the bakers' street, until all the bread in the city was gone."

According to another version (Jer. 38:1 ff.), the officials heard of Jeremiah's prophecy that Jerusalem would be given into the hands of the king of Babylon, that its inhabitants would die of famine, sword and pestilence, and that only "whoever surrenders to the Chaldeans" would be saved from death.[36] They therefore said to the king: "Let that man be put to death, for he disheartens (lit., 'weakens the hands of') the soldiers and all the people who are left in the city by speaking such things to them. That man is not seeking the welfare of this people, but their harm." This accusation was so serious that even the king himself could not prevent appropriate measures being taken against the prophet and he turned him over to the officials. Jeremiah was thrown into a pit in the prison compound, where he was in danger of death, and saved only thanks to Zedekiah's order to remove him from the pit. Jeremiah remained in the prison compound "until the day Jerusalem was captured."

There are several striking parallels in the key elements of these versions: in both cases the prophet is accused of a very severe crime in face of the enemy (desertion to the Chaldeans in the one and disheartening the people

36 In Jer. 21:9, the prophet says almost the same thing: "… but he who leaves and deserts to the Chaldeans (והיוצא ונפל על הכשדים) who are besieging you shall live; he shall at least gain his life (lit., "he shall have his life as booty"). Both נפל אל/על and יצא אל indicate surrender to the enemy. However, it seems that the expression נפל אל/על indicates voluntary surrender with intent to desert, while יצא אל indicates surrender by necessity, due to enemy superiority and the recognition that surrender would minimize the consequences. Compare Jeremiah's response to Zedekiah: "If you surrender (יצא תצא אל) to the officers of the King of Babylon, your life will be spared and the city will not be burned down. But if you do not surrender (ואם לא תצא) to the officers of the King of Babylon, this city will be delivered into the hands of the Chaldeans, who will burn it down, and you will not escape from them" (Jer. 38:17–18); compare also Rabshakeh's comments on the impossibility of Jerusalem's deliverance from Sennacherib and his proposal: "Thus says the King of Assyria: make your peace with me and come out to me (וצאו אלי)… so that you may live and not die" (2 Kgs. 18:28–35); cf., 1 Sam. 11:3; 1 Kgs. 20:31. The phrasing in Jer. 21:9, "he who leaves and deserts to the Chaldeans who are besieging you shall live and shall have his life as booty," suggests that Jeremiah was indeed advocating desertion.

in the other), he is given over to the officials and imprisoned under harsh conditions to the point of endangering his life and finally saved thanks to Zedekiah, who grants him special treatment (possibly because of his status as a prophet), and orders his transfer to the prison compound, where the conditions are more tolerable. Consequently, the two stories (Jer. 37:5–21 and 38:1 ff) should be regarded as versions of the same event – Jeremiah's imprisonment in the later stages of the siege on Jerusalem – one version composed in Babylonia and the other in the local Judean environment.[37]

The guidelines given by Aeneas the Tactician reveal a noteworthy phenomenon: while he emphasizes strict supervision of elements whose loyalty is questionable and recommends extensive preventive measures to ensure they do not endanger the city's standing power, there is no sign of urgency in taking severe measures against these elements, so long as crimes have not actually been committed. This moderation is to be explained not only on the basis of the political and social regime that was dominant in the Greek cities (an approach that was probably appropriate in the ancient Near East as well), but also from the recognition that strict behavior of the authorities toward various groups during an emergency may be rash and harm them and their successors in the future, after the siege.

2. Treatment of the Wounded

The fear of injury in battle is felt by every soldier, and the knowledge that suitable treatment will be available to the wounded is an important factor contributing to one's readiness to risk oneself in battle. It is therefore reasonable to assume that leaders who were concerned with the morale

37 See discussion in Ephʿal (*op. cit.*, n. 35), 20. A different version appears in Jer. 32:1–5, according to which Jeremiah was imprisoned by Zedekiah in the prison compound because he had prophesied during the siege that the city would fall into the hands of the King of Babylon and that Zedekiah would be given over to him and taken to Babylonia. The figure of Zedekiah, who punishes the prophet for his harsh words, is inconsistent with the image of him that emerges from Jer. 37:17–21; 38:7–27, according to which the king sought the word of God through Jeremiah, accepted the prophet's response with resignation (but refused to surrender to the Babylonian officers because "I am afraid of the Judeans who have defected to the Chaldeans, that they (the Chaldeans) might hand me over to them to abuse me", 38:19), and showed concern for the prophet's well-being.

of the combatants and the rest of the population would act to ensure such treatment.[38]

Although fortifications provided some protection for the defenders, there is no doubt that some were injured by enemy fire or exposed to danger during counter attacks and sorties outside of the city.[39] The use of artillery, when it became established in siege warfare, also increased the danger to the non-combatant population.[40] Locals who were injured were probably cared for by their families; the situation was more difficult for the foreign wounded.

A document recorded on a bronze tablet in the Cypriote syllabic script, found in Idalium, Cyprus, sheds light on the initiatives taken by the authorities to assure medical care to all wounded. The tablet records a contract between King Stasikypros and the city of Idalium, and the physician, Onasilos son of Onasikypros.[41] In this document, "the king and the city" agree to give the physician and his brothers a talent of silver (plus additional silver in the amount of 4 "hatchets" + 2 double minas(?)) or, alternatively, land and orchards from "the royal lands," as outlined in the contract, as chosen by the physician and his brothers, in exchange for the treatment of anyone wounded in battle, without further payment. The contract was deposited in the temple of Athena in Idalium, and its closing section states that whoever attempts to violate the contractual obligations will be considered a blasphemer.

The document notes the year in which it was written – the eponym year of Philokypros son of Onasagoras – but we have no information by

38 For a request to bring a physician (^{lú}*asûm*) to the city of Nahur in preparation for a siege, see ARM II 127.

39 Reliefs depicting battle over cities show some of the defenders, pierced with spears or arrows, falling from the wall outwards; see, for example, Yadin, *Art of Warfare*, 228, 229, 347 (Egypt); 406, 448, 449 (Assyria). It seems probable that in reality, most of those injured on the wall fell on the wall itself or inwards, towards the city. The depiction of falling outwards is probably derived from the technical/artistic constraint, as they cannot be shown falling behind the solid wall.

40 See above, pp. 103–105.

41 See Masson, *Les inscriptions chypriotes syllabiques*, no. 217, pp. 235–244 (and references therein).

which to determine the precise date. The opening sentence mentions the circumstances under which the contract was made. Since the publication of this document (over a century ago), the accepted reading of this sentence has been: "When the Medes, together with the people of Kitium, laid siege to Idalium…" Accordingly, it was viewed as evidence of a Persian siege on Idalium between 478–470 BC. Rosén has questioned this reading and its interpretation.[42] In his opinion, the text does not refer to "siege" or to "Medes" (in addition, he reads "the Kitian" rather than "the Kitians"). Rosén argues that the contract is not associated with any particular siege; rather, it is simply a work contract between the "state" (i.e., "the king and the city") and the physician and his brothers, according to which the latter would take care of anyone injured in battle during the year specified in the document. Rosén's version provides a wider interpretation than the accepted reading: rather than a specific siege under well-defined circumstances, the contract designates a broader situation, referring to any battle that may take place during the year of the contract (evidently including siege battles as well). If this interpretation is correct,[43] it is reasonable to assume that such contracts were common practice and periodically renewed. Either way, the contents of the document in both interpretations reveal a factor of crucial importance regarding the physical welfare of the besieged population and its morale.

3. The Liberation of Slaves

Jer. 34:8–22 mentions the liberation of Hebrew slaves in Zedekiah's reign and their later repossession, probably during the lull in Nebuchadnezzar's siege of Jerusalem, following reports that the Egyptian army was coming to the aid of Judah (see vv. 21–22). According to verses 8–10, 15, 18–19, the slaves were freed after a covenant was made in the Temple with the active participation of the king, "the officers of Judah and Jerusalem, the officials, the priests, and all the people of the land." The phrasing of the text explaining the reason for the slaves' release, as given in vv. 8–9, 14,

42 See: Rosén, in *Symposion 1977*, 9–32.

43 For a response to Rosén's remarks and a defense of the common view, see: Masson, *Bulletin de la Société de Linguistique* 88/1 (1983), 261–267.

is clearly related to the verses in Deuteronomy 15:1–18 regarding the laws of the sabbatical year and the laws of the Hebrew slave, and bears a partial relation to Exodus 21:2 and Leviticus 25:39–43.[44] This similarity has led scholars to suggest that the event occurred in a sabbatical or jubilee year, identified as 588/7 BC, which required the release of slaves in accordance with biblical law.[45] The opinion that the text refers to an event that occurred in 588/7, during the siege of Jerusalem,[46] relies on the assumption that the repossession of slaves took place once the Babylonian siege was lifted.

Under this assumption, it may be argued that the liberation of the slaves stemmed from strictly military and economic considerations, and not necessarily from the desire to fulfill the Torah laws of slavery; Jer. 34:14 even seems to imply that the law requiring the liberation of Hebrew slaves was not practiced for a certain period prior to the event described. In other words, the slaves were freed because their maintenance was a burden to their masters, or perhaps because it enabled the authorities to enlist the participation of the freed male slaves in the city's defense[47] (note that, in principle, Hebrew slaves were only indentured servants, that is, citizens of

44 A comparison of the biblical slavery laws with Jer. 34:14 has led some scholars to suggest alterations in the text, while others have suggested detaching Jer. 34:12–22 from historical reality and seeing it as a midrash (see: Carroll, *The Book of Jeremiah*, 646–650). In my opinion, the text can be understood without recourse to such a far-reaching approach; the textual corrections suggested are not germane to our discussion and need not be brought here.

45 David, *OTS* 5 (1948), 63–79; Sarna, in *Orient and Occident*, 143–149; Lemche, *VT* 26 (1976), 38–59.

46 This date implies that the siege of Jerusalem began on 10 Tebeth 588 BC and continued for 30 months. Others have argued that the siege began in 587 BC and therefore lasted only 18 months; see Galil, *The Chronology of the Kings of Israel and Judah,* 108–126.

47 On the liberation of slaves in the city of Upi during the general mobilization that was conducted by Hammurabi in the course of his war against the Elamites, see ARM XXVI 363:9–15; see also Anbar, *ZAW* 111(1999), 253–255. On the positioning of liberated slaves against the enemy, in preparation for attack on the besieged city, see Diodorus, 17.11.2 (re Alexander's attack on Thebes). On the decision of the Spartans to free helots in consideration of their participation in military service in times of war and shortage or warriors, see Thucydides 5.34; Xenophon, *Hellenica* 6.5.28 (421 and 370/69 BC, respectively).

Judah whose term of servitude was no more than six years).[48] The same approach is evident in Aeneas the Tactician (14.1), who recommends the reduction or cancellation of citizens' (πολῖται) debts, in order to ensure their sympathies and loyalty, thus strengthening morale and solidarity.

However, this assumption is difficult to accept. It is unlikely that the liberation of Hebrew slaves during the siege, officially confirmed in the Temple ceremony, was done with the intention to repossess the same individuals afterwards. There is no law permitting the seizure of slaves against their will after they were freed. Even if we accept some of Jeremiah's comments about the atmosphere of corruption that was predominant in the kingdom of Judah in its final days, it is highly doubtful that this kingdom, like any organized society, could function with the degree of legal anarchy implied by this hypothesis.

An alternative hypothesis allows us to maintain the historicity of Jer. 34:8–22 and to connect the text to the biblical laws of slavery. According to this hypothetical sequence of events, the slaves were set free in 588/7 BC, under the law of the sabbatical or jubilee year; however, economic conditions following the Babylonian attack on Judah made it difficult for them to survive independently and their masters were unable to provide them with a means of livelihood (see Deut. 15:13–14). Under these circumstances, a significant number of freed slaves may have chosen to voluntarily re-enslave themselves for a set period of time (not permanent enslavement, as implied by Deut. 15:16–17 and Exod. 21:5–6), in anticipation of economic recovery in the following year, which would have enabled them to support themselves as free people. The optimism which prevailed in Judah during the cessation of the siege (before its temporary nature became apparent) allowed the liberated slaves and their former masters to renew the connection between them[49] – and it was this that Jeremiah protested. Thus,

48 Jeremiah only refers to Hebrew slaves. It follows, therefore, that the gentile slaves were not set free (which was not required by law).

49 In light of this approach and given the economic conditions noted, it seems reasonable to assume that this relationship was not that of salaried employment involving all the conditions legally associated with such a connection, since it was probably difficult for employers to pay their freed slaves for their work. Therefore, it seems that it was in the interest of both parties to temporarily continue the previous state of affairs.

we argue that forced re-enslavement of liberated slaves was impossible in the kingdom of Judah.

BIBLIOGRAPHY

Abrahami, Ph., 'La circulation militaire dans les textes de Mari: la question des effectifs,' in D. Charpin and F. Joannès (eds.), *La circulation des biens, des personnes et des idées dans le Proche-Orient ancien*, Paris 1992, 157–166.

Adamson, P. B., 'Problems over Storing Food in the Ancient Near East,' *WO* 16 (1985), 5–15.

Albenda, P., 'A Syro-Palestinian(?) City on a Ninth Century B.C. Assyrian Relief,' *BASOR* 206 (1972), 42–48.

Anbar, M. 'La libération des esclaves en temps de guerre: Jer 34 et ARM XXVI, 363,' *ZAW* 111 (1999), 253–255.

Arnaud, D., 'Les textes d'Emar et la chronologie de la fin du Bronze récent,' *Syria* 52 (1975), 87–92.

—, *Recherches au pays d'Aštata. Emar* VI/1-4, Paris 1985–1987.

—, 'La Syrie du moyen-Euphrate sous le protectorat hittite: contrats de droit privé,' *AuOr* 5 (1987), 211–241.

—, *Textes syriens de l'âge du bronze récent*, Barcelona 1991.

—, 'Tablettes de genres divers du Moyen-Euphrate,' *SMEA* 30 (1992), 195–245.

Attridge, H. W. and Oden, R. A., *Philo of Biblos: The Phoenician History*, Washington 1981.

Avishur, Y., *Studies in Hebrew and Ugaritic Psalms*, Jerusalem 1994.

Bachi, G. and Ben-Dov, M., 'Area D,' *Ashdod* II–III, Jerusalem 1971 (= ᶜ*Atiqot*, English Series 9-10), 86–124.

Balkan, K., *Letter of King Anum-ḫirbi of Mama to King Warshama of Kanish*, Ankara 1957.

Bar-Kochva, B., *Judas Maccabaeus: The Jewish Struggle against the Seleucids*, Cambridge 1989.

Barnett, R. D., 'The Amathus Shield-boss Rediscovered and the Amathus Bowl Reconsidered,' *RDAC* 1977, 157–169.

Barnett, R. D. and Falkner, M., *The Sculptures of Aššur-naṣir-apli II (883–859 B.C.), Tiglath-pileser III (747–727 B.C.), Esarhaddon (681–669 B.C.) from the Central and South-West Palaces at Nimrud*, London 1962.

Barnett, R. D. and Lorenzini, A., *Assyrian Sculpture in the British Museum*, Toronto 1975.

Beaulieu, P.-A., 'The Fourth Year of Hostilities in the Land,' *BaM* 28 (1997), 367–393.

Beckman, G., 'The Siege of Uršu Text (*CTH* 7), and Old Hittite Historiography,' *JCS* 47 (1995), 23–34.

—., *Texts from the Vicinity of Emar in the Collection of Jonathan Rosen*, Padova 1996.

Bilgiç, E., 'Die wichtigsten Ausdrücke über Darlehen in den Keilschrifttexten,' *Ankara Üniversitesi Dil ve Tarih-Coğrafya Fakültesi Dergisi* 5 (1947), 447–454.

Billerbeck, A. and Delitzsch, F., *Die Palasttore Salmanassars II. von Balawat*, Leipzig 1908.

Biran, A. and Naveh, J., 'An Aramaic Stele Fragment from Tel Dan,' *IEJ* 43 (1993), 81–98.

Birot, M., *Correpondance des gouverneurs de Qaṭṭunân*, Paris 1993.

Blinkenberg, Chr., *Lindos, Fouille de l'Acropole 1902–1914*, II. *Inscriptions*, Tome I, Berlin-Copenhagen 1941.

Boardman, J. (ed.), *Cambridge Ancient History, Plates to Volume III*, Cambridge 1984.

Borger, R., *Die Inschriften Asarhaddons, Königs von Assyrien*, Graz 1956.

—, 'Mesopotamien in den Jahren 629–621 v. Chr.,' *WZKM* 55 (1959), 62–76.

—, 'Zu den Asarhaddon-Verträgen aus Nimrud,' *ZA* 54 (1961), 173–196.

—, 'Der Aufsieg des neubabylonischen Reiches,' *JCS* 19 (1965), 59–78.

—, 'Gott Marduk und Gott-König Šulgi als Propheten. Zwei prophetische Texte,' *BiOr* 28 (1971), 3–24.

—, 'Review on *CAD* vol. K,' *BiOr* 32 (1975), 70–72.

—, *Beiträge zum Inschriftenwerk Assurbanipals*, Wiesbaden 1996.

Botta, P. E. and Flandin, E., *Monument de Ninive*, I/II, Paris 1849.

Brinkman, J. A., 'Sennacherib's Babylonian Problem: An Interpretation,' *JCS* 25 (1973), 89–95.

—, *Prelude to Empire: Babylonian Society and Politics, 747–626 B.C.*, Philadelphia 1984.

Brinkman, J. A. and Kennedy, D. A., 'Documentary Evidence for the Economic Base of Early Neo-Babylonian Society: A Survey of Dated Babylonian Economic Texts, 721-626 B.C.' *JCS* 35 (1983), 1-90.

—, 'Supplement to the Survey of Dated Neo-Babylonian Texts, 721-626 B.C. (*JCS* 35 [1983], 1-90),' *JCS* 38 (1986), 99-106.

Broshi, M. and Finkelstein, I., 'The Population of Palestine in the Iron Age II,' *BASOR* 287 (1992), 47-60.

Brown, Sh., *Late Carthaginian Child Sacrifice and Sacrificial Monuments in their Mediterranean Context*, Sheffield 1991.

Carrol, R. P., *Jeremiah: A Commentary*, Philadelphia 1986.

Charpin, D., 'Donées nouvelles sur la poliorcétique à l'époque paléo-Babylonienne,' *MARI* 7 (1993), 193-203.

Charpin, D., Joannès, F., Lackenbacher, S. and Lafont, B., *Archives épistolaires de Mari* I/1-2, Paris 1988.

Clark, C. and Haswell, M., *The Economics of Subsistence Agriculture*, London 1967[3], 1970[4].

Cogan, M. and Tadmor, H., 'Ashurbanipal's Conquest of Babylon: The First Official Report – Prism K,' *Orientalia* 50 (1981), 229-240.

—, *II Kings*, Anchor Bible, Garden City, N.Y., 1988.

Cohen, H. R. (Chaim), *Biblical Hapax Legomena in the Light of Akkadian and Ugaritic*, Missoula, Mont. 1978.

Cohen, Ch., 'Neo-Assyrian Elements in the First Speech of the Biblical Rab-Šāqê,' *IOS* 9 (1979), 32-48.

Cole, S. W., 'The Destruction of Orchards in Assyrian Warfare,' in S. Parpola and R. Whiting (eds.), *Assyria 1995*, Helsinki 1997, 29-40.

Cook, J. M., 'Old Smyrna, 1948-1951,' *BSA* 53-54 (1958-1959), 1-34.

Cooper, J. S., 'Apodictic Death and the Historicity of "Historical" Omens,' in B. Alster (ed.), *Death in Mesopotamia*, XXVIe RAI (= Mesopotamia 8), Copenhagen 1980, 99-105.

—, *The Curse of Agade*, Baltimore-London 1983.

Crété, L., *La vie quotidienne à La Rochelle au temps du grand siège:*

1627-1628, Paris 1987.

Da Riva, R., 'BM 114354: un nuevo texto *edêl bâbi* redactado en Uruk durante la guerra entre Nabopolasar y Sîn-šar-iškun, '*AuOr* 20 (2002), 249-251.

Dalley, S. M., 'Neo-Assyrian Tablets from Til-Barsib,' *Abr-Naharain* 34 (1996-1997), 66-99.

Dalley, S. and Teissier, B., 'Tablets from the Vicinity of Emar and Elsewhere,' *Iraq* 54 (1992), 83-112.

Dandamaev, M. A., *Slavery in Babylonia*, DeKalb 1984.

Darnell, J. C., 'Two Sieges in the Aethiopic Stelae,' in D. Mandel and U. Claudi (eds.), *Ägypten im afro-orientalischen Kontext: Aufsätze zur Archäologie, Geschichte und Sprache eines unbegrenzten Raumes (Gedenkschrift P. Behrens*), Köln 1991, 73-93.

David, M., 'The Manumission of Slaves under Zedekiah,' *OTS* 5 (1948), 63-79.

Day, J., *Molech: A God of Human Sacrifice in the Old Testament*, Cambridge 1989.

Deller, K., 'Getreidekursangaben in neuassyrischen Rechtsurkunden,' *Orientalia* 33 (1964), 257-261.

Diels, H. and Schramm, E., *Excerpte aus Philons Mechanik B. VII und VIII*, Berlin 1920.

Dietrich, M., ''Die akkadischen Texte der Archive und Bibliotheken von Emar,' *UF* 22 (1990), 25-48.

Dietrich, M., Loretz, O. and Sanmartin, J., *The Cuneiform Alphabetic Texts from Ugarit, Ras Ibn Hani and Other Places*, Münster 1995.

Donner, H. and Röllig, W., *Kanaanäische und armäische Inschriften*, Wiesbaden 1964-1966.

Dossin, G., *Correspondance de Šamši-Addu et ses fils*, Paris 1950.

Dothan, M., *Ashdod* II-III, Jerusalem 1971 (= ᶜ*Atiqot*, English Series 9-10).

Dougherty, R. Ph., *Records from Erech, Time of Nabonidus (555-538 B.C.)*, New Haven 1920.

—, *The Shirkûtu of Babylonian Deities*, New Haven 1923.

—, *Archives from Erech, Time of Nebuchadrezzar and Nabonidus*, New Haven 1923.

Driver, G. R., 'Ezekiel: Linguistic and Textual Problems,' *Biblica* 35 (1954), 145–159.

Dubberstein, W. H., 'Comparative Prices in Later Babylonia (625–400 B.C.),' *AJSL* 56 (1939), 20–43.

Ebeling, E., *Keilschrifttexte aus Assur religiösen Inhalts*, I–II, Leipzig 1915–1923.

—, 'Die Rüstung eines babylonischen Panzerreiters nach einem Vertrag aus der Zeit Darius II,' *ZA* 36 (1952), 203–213.

Ehrlich, A. B., *Miqrâ ki-Peschuto (Die Schrift nach ihrem Wortlaut)*, Berlin 1899.

Eidem, J. and Laessøe J., *The Shemshara Archive*, 1. *The Letters*, Copenhagen 2001.

Ellis, M. deJ., 'Neo-Babylonian Texts in the Yale Babylonian Collection,' *JCS* 36 (1984), 1–63.

Ephᶜal, I., 'On Warfare and Military Control in the Ancient Near Eastern Empires: A Research Outline,' in H. Tadmor and M. Weinfeld (eds.), *History, Historiography and Interpretation: Studies in Biblical and Cuneiform Literatures*, Jerusalem 1983, 88–106.

—, 'The Assyrian Siege Ramp at Lachish: Military and Lexical Aspects,' *TA* 11 (1984), 60–70.

—, '"You Are Defecting to the Chaldeans" (Jer. 37:13),' *EI* 24 (1993), 18–22 (Hebrew).

—, 'On Common Literary Expressions of the Ancient Semites,' in Ch. Cohen, A. Hurvitz and Sh. M. Paul (eds.), *Sefer Moshe* (M. Weinfeld Jubilee Volume), Winona Lake 2004, 25–32.

—, 'Esarhaddon, Egypt, and Shubria: Politics and Propaganda,' *JCS* 57 (2005), 99–111.

Ephᶜal, I. and Tadmor, H., 'Observations on Two Inscriptions of Esarhaddon: Prism Nineveh A and the Letter to the God,' in Y. Amit, E. Ben Zvi, I. Finkelstein and O. Lipschits (eds.), *Essays on Ancient Israel and Its Near Eastern Context* (Festschrift N. Naʾaman), Winona Lake 2006, 155–170.

Erdmann, E., in F. G. Maier (ed.), *Ausgrabungen in Alt-Paphos auf Cypern*, Heft I: *Nordosttor und persische Belagerungsrampe in Alt-Paphos*, Konstanz 1977.

Fales, F. M., 'Notes on the Royal Family of Emar,' in D. Charpin and F. Joannès (eds.), *Marchands, diplomates et empereurs: Études sur la civilisation mésopotamienne offertes à P. Garelli*, Paris 1991, 81–90.

Fales, F. M. and Jakob-Rost, L., 'Neo-Assyrian Texts from Assur: Private Archives in the Vorderasiatisches Museum of Berlin, Part I,' *SAAB* 5 (1991), 3–157.

Falkner, M., 'Die Eponymen der spätassyrischen Zeit,' *AfO* 17 (1954–56), 100–120.

Farber, H., 'A Price and Wage Study for Northern Babylonia during the Old Babylonian Period,' *JESHO* 21 (1978), 1–51.

Foster, B. R., 'The Siege of Armanum,' *JANES* 14 (1982), 27–36.

—, *Before the Muses*, I-II, Bethesda, MD 1993.

Frame, G., *Babylonia 689–627 B.C.: A Political History*, Istanbul 1992.

—, 'A Siege Document from Babylon Dating to 649 B.C.,' *JCS* 51 (1999), 101–106.

Fuchs, A. and Parpola, S., *The Correspondence of Sargon II, Part III*, Helsinki 2001.

Galil, G., *The Chronology of the Kings of Israel and Judah*, Leiden-New York-Köln 1996.

Gevirtz, S., 'On Canaanite Rhetoric: The Evidence of the Amarna Letters from Tyre,' *Orientalia* 42 (1973), 162–177.

Glasner, J.-J., 'Narām-Sīn poliorcète: les avatars d'une sentence divinatoire,' *RA* 77 (1983), 3–10.

Goedicke, H., 'The Capture of Joppa,' *Chronique d'Égypte* 43 (1968), 219–233.

—, 'Some Remarks Concerning the Inscriptions of Ahmose, Son of Ebana,' *JARCE* 11 (1974), 31–41.

—, 'The End of the Hyksos in Egypt,' in L. H. Lesko (ed.), *Egyptological Studies in Honor of R. A. Parker*, Hanover, NH 1986, 37–47.

Goetze, A., *Old Babylonian Omen Texts*, New Haven-London 1947.

Goodnick Westenholz, J., *Cuneiform Inscriptions in the Collection of the Bible Lands Museum, Jerusalem: The Emar Tablets*, Groningen 2000.

Grayson, A. K., *Assyrian and Babylonian Chronicles*, Locust Valley, N.Y. 1975.

—, *Assyrian Rulers of the Third and Second Millennia BC (to 1115 BC)*,

Toronto 1987.

—, *Assyrian Rulers of the Early First Millennium BC, I (1114–859 BC)*, Toronto 1991.

—, *Assyrian Rulers of the Early First Millennium BC, II (858–745 BC)*, Toronto 1996.

Greenberg, M., *Ezekiel, 1–20,* Anchor Bible, Garden City, N.Y. 1983.

Greenfield, J. C., *ᶜAl Kanfei Yonah* (Collected Studies of Jonas C. Greenfield on Semitic Philology), Leiden-Jerusalem 2001.

Grimal, N-C., *La stèle triomphale de Pi(ᶜankh)y au Musée du Caire*, Cairo 1981.

Gurney, O. R., 'The Sultantepe Tablets: IV. The Cuthaen Legend of Naran-Sin,' *AnSt* 5 (1955), 93–113.

—, 'The Sultantepe Tablets: VII. The Myth of Nergal and Ereshkigal,' *AnSt* 10 (1960), 105–131.

Güterbock, H. G., 'Die historische Tradition und ihre literarische Gestaltung bei Babyloniern und Hethitern bis 1200, Zweiter Teil: Hethiter,' *ZA* 44 NF 10 (1938), 45–149.

Haas, N., 'Anthropological Observations on the Skeletal Remains found in Area D (1962–1963),' *Ashdod* II–III, Jerusalem 1971 (= *ᶜAtiqot* 9–10), 212–214.

Hallo, W. W., 'Proverbs Quoted in Epic,' in T. Abusch et al. (eds.), *Lingering Over Words: Studies in Ancient Near Eastern Literature in Honor of W. L. Moran*, Atlanta, GA 1990, 203–217.

Hammond, N. G. L., *Three Historians of Alexander the Great*, Cambridge 1983.

Handcock, P. S. P., *Cuneiform Texts from the Babylonian Tablets in the British Museum*, London 1911.

Harper, R. F., *Assyrian and Babylonian Letters Belonging to the K(ouyunjik) Collection(s) of the British Museum*, 1–14, Chicago 1892–1914.

Hawkins, J. D., 'Royal Statements of Ideal Prices: Assyrian, Babylonian, and Hittite,' in J. V. Canby et al. (eds.), *Ancient Anatolia: Aspects of Change and Cultural Development, Essays in Honor of M. J. Mellink*, Madison, Wis. 1986, 93–102.

Heidel, A., 'The Octagonal Sennacherib Prism in the Iraq Museum,' *Sumer* 9 (1953), 117–187.

Held, M., 'Studies in Comparative Semitic Lexicography: I. Akkadian *zē summāti* = Hebrew *ḥ^a^rē yōnīm*' in *Studies in Honor of B. Landsberger*, Chicago 1965, 395–398.

Heltzer, M., '*Dimtu-gt-pýrgos*: An Essay about the Non-etymological Sense of these Terms,' *JNSL* 7 (1979), 31–35.

—, 'The Persepolis Documents, the Lindos Chronicle and the Book of Judith,' *La Parola del Passato* 44 (1989), 81–101.

Herzog, Z., 'Settlement and Fortification Planning in the Iron Age,' in A. Kempinski et al. (eds.), *The Architecture of Ancient Israel from the Prehistoric to the Persian Periods*, Jerusalem 1992, 231–274.

Hillers, D. R., *Treaty-Curses and the Old Testament Prophets*, Rome 1964.

Houtman, C., 'What did Jacob see in his Dream at Bethel?', *VT* 27 (1977), 337–351.

Houwink ten Cate, Ph. H. J., 'The History of Warfare according to Hittite Sources: The Annals of Hattusilis I.' *Anatolica* 10 (1983), 91–109; 11 (1984), 47–83.

Hunger, H., 'Das Archiv des Nabû-ušallim,' *BaM* 5 (1970), 193–304.

Hunter, L. W., *Aeneas on Siegecraft*, Oxford 1927.

The Illinois Greek Club (members of), *Aeneas Tacticus; Asclepiodotus; Onasander*, London–New York 1923.

Ismail, B. Kh., 'Eine Siegesstelle des Königs Daduša von Ešnunna,' in W. Meid and H. Trenkwalder (eds.), *Im Bannkreis des Alten Orients* (Festschrift Karl Oberhuber), Innsbruck 1986, 105–108.

—, 'Dādušas Siegesstele IM 95200 aus Ešnunna. Die Inschrift,' *BaM* 34 (2003), 129–163.

Izreʾel, Sh. and Singer, I., *The General's Letter from Ugarit*, Tel Aviv 1990.

Jacoby, R., 'The Representation and Identification of Cities on Assyrian Reliefs,' *IEJ* 41 (1991), 112–131.

Jaroš-Deckert, B., *Das Grab des Jnj-jtj.f: Die Wandmalereien der XI. Dynastie (Grabungen in Asasif 1963–1970*, Bd. V), Mainz am Rhein 1984.

Jean, Ch.-F., *Lettres diverses*, Paris 1950.

Kaiser, O. et al. (eds.), *Texte aus der Umwelt des Alten Testaments*,

Gütersloh 1982–.

Keel, O., 'Kanaanäische Sühneriten auf ägyptischen Tempelreliefs,' *VT* 25 (1975), 413–469.

Keita, S. O. Y., 'An Analysis of Crania from Tell-Duweir Using Multiple Discriminant Functions,' *American Journal of Physical Anthropology* 75 (1988), 375–390.

Keith, A., 'The Men of Lachish,' *PEQ* 72 (1940), 7–12.

Kennedy, D. A., 'Documentary Evidence for the Economic Base of Early Neo-Babylonian Society, Part II: A Survey of Babylonian Texts, 625–605 B.C.,' *JCS* 38 (1986), 172–244.

Klauber, E. G., *Politisch-religiöse Texte aus der Sargonidenzeit*, Leipzig 1913.

Klengel, H., '"Hungerjahre" in Ḫatti,' *AOF* 1 (1974), 165–174.

Knudsen, E. E., 'Fragments of Historical Texts from Nimrud – II,' *Iraq* 29 (1967), 49–69.

Knudtzon, J. A., *Assyrische Gebete an den Sonnengott*, Band II, Leipzig 1893.

—, *Die El-Amarna-Tafeln*, Leipzig 1915.

Kramer, S. N., *Lamentation Over the Destruction of Ur*, Chicago 1940.

Kümmel, H. M., *Familie, Beruf und Amt im spätbabylonischen Uruk*, Berlin 1979.

Kutscher, E. Y., '*Bešulei hamilon hamiqraʾi,*' *Lešonenu* 27–28 (1963–1964), 183–188 (Hebrew).

Kwasman, Th. and Parpola, S., *Legal Transactions of the Royal Court of Nineveh, Part I: Tiglath-Pileser III through Esarhaddon*, Helsinki 1991.

Labat, R., *Textes litéraires de Suse*, Paris 1974.

Lassøe, J., 'IM 62100: A Letter from Tell Shemshara,' in *Studies in Honor of B. Landsberger*, Chicago 1965, 189–196.

Lambert, W. G., 'The "tamītu" Texts,' in *La divination en Mésopotamie ancienne et dans le régions* (XIVe RAI), Paris 1966, 119–123.

—, 'Questions Addressed to the Babylonian Oracles: The *tamītu* Texts,' in J.-G. Heintz (ed.), *Oracles et prophéties dans l'antiquité*, Strasburg 1997, 85–98.

Lambert, W. G. and Millard, A. R., *Atra-ḫasīs, The Babylonian Story of the*

Flood, Oxford 1969.

Langdon, S., *Die neubabylonischen Königsinschriften*, Leipzig 1912.

Larsen, M. T., 'Unusual Eponymy-dating from Mari and Assyria,' *RA* 68 (1974), 15–24.

Layard, A. H., *The Monuments of Nineveh*, I, London 1849.

Leichty, E., *The Omen Series šumma izbu*, Locust Valley 1970.

—, 'Esarhaddon's 'Letter to the Gods',' in M. Cogan and I. Ephᶜal (eds.), *Ah Assyria... (Studies in Assyrian History and Ancient Near Eastern Historiography presented to H. Tadmor*), Jerusalem 1991, 52–57.

Lemche, N. P., 'The Manumission of Slaves – The Fallow Year – The Sabbatical Year – The Jobel Year.' *VT* 26 (1976), 38–59.

Lichtheim, M., *Ancient Egyptian Literature*, II. *The New Kingdom* (1976), III. *The Late Kingdom* (1980), Berkeley-Los Angeles-London.

Liddell, H. G. and Scott, R., *A Greek-English Lexicon*, Oxford 1940.

Lie, A. G., *The Inscriptions of Sargon II, King of Assyria*, Part I: *The Annals*, Paris 1929.

Liphschitz, N., '*Ceratonia siliqua* in Israel: An Ancient Element or a Newcomer?' *Israel Journal of Botany* 36 (1987), 191–197.

Lipiński, E., '*Nešek* and *Tarbît* in the Light of Epigraphic Evidence,' *OLP* 10 (1979), 133–141.

Liverani, M., '*Kitru, katāru*,' *Mesopotamia* 17 (1982), 43–66.

—, *Prestige and Interest: International Relations in the Near East ca. 1600–1100 B.C.*, Padova 1990.

Livingstone, A., *Court Poetry and Literary Miscellanea*, Helsinki 1989.

Loewenstamm, S. E., *From Babylon to Canaan* (Studies in the Bible and its Oriental Background), Jerusalem 1992.

Loewenstamm, S. E. and Blau, J., *Thesaurus of the Language of the Bible*, II, Jerusalem 1959.

Lohfink, N., 'Deuteronomy 6:24 לְחַיֹּתֵנוּ "To Maintain Us",' in M. Fishbane et al. (eds.), *Shaᶜarei Talmon* (*Studies in the Bible, Qumran, and the Ancient Near East Presented to Sh. Talmon*), Winona Lake 1992, 111–119.

Luckenbill, D. D., *The Annals of Sennacherib*, Chicago 1924.

Lutz, H. F., 'An Agreement Between a Babylonian Feudal Lord and His Retainer in the Reign of Darius II,' *UCP* 9/III, Berkeley 1928,

269–277.

MacGinnis, J., 'Neo-Babylonian Prebend Texts from the British Museum,' *AfO* 38–39 (1991-1992), 74–100.

Machinist, P. B., *The Epic of Tukulti-Ninurta I: A Study in Middle Assyrian Literature*, PhD dissertation, Yale University 1978.

Maier, F. G., 'Ausgrabungen in Alt-Paphos, 1950–1971,' *Chiron* 2 (1972), 17–35.

—, *Archäologie und Geschichte. Ausgrabungen in Alt-Paphos*, Konstanz 1973.

Maier, F. G. and Karageorghis, V., *Paphos: History and Archaeology*, Nicosia 1984.

Maier, F. G. and Wartburg, M.-L., 'Reconstructing History from the Earth, c. 2800 B.C.–1600 A.D.: Excavation at Palaepaphos, 1966–1984,' in V. Karageorghis (ed.), *Archaeology in Cyprus 1960–1985*, Nicosia 1985, 141–172.

Malamat, A., 'The Last Kings of Judah and the Fall of Jerusalem. An Historical–Chronological Study,' *IEJ* 18 (1968), 149–159.

—, 'Conquest of Canaan: Israelite Conduct of War According to Biblical Tradition,' *History of Biblical Israel*, Leiden-Boston-Köln 2001, 68–96.

Margalit, B., 'A Ugaritic Prayer for a City Under Siege,' *Proceedings of the Seventh World Congress of Jewish Studies: Studies in the Bible and the Ancient Near East*, Jerusalem 1981, 63–83 (Hebrew).

Marsden, E. W., *Greek and Roman Artillery. Historical Development*. Oxford 1969.

Masson, O., *Les inscriptions chypriotes syllabiques*, Paris 1961.

—, 'Remarques sur quelques passages de la tablette chypriote d'Idalion (*ICS* 217),' *Bulletin de la Société de Linguistique* 88/1 (1983), 261–281.

McEwan, G. J. P., *Late Babylonian Texts in the Ashmolean Museum*, Oxford 1984.

Meissner, B., 'Warenpreise in Babylonien,' *Abhandlungen der preussischen Akademie der Wissenschaften*, Phil.-hist. Klasse, Berlin 1936, No. 1.

Moran, W. L., *The Amarna Letters*, Baltimore-London 1992.

Mosca, P. G., *Child Sacrifice in Canaanite and Israelite Religion*, PhD dissertation, Harvard University 1975.

Na᾿aman, N., 'Sennacherib's "Letter to God" on His Campaign to Judah,' *BASOR* 214 (1974), 25–39.

—, 'Campaigns of the Assyrian Kings to Judah in the Light of a New Assyrian Document,' *Shnaton* 2 (1977), 164–180 (Hebrew).

—, 'The Historical Background to the Conquest of Samaria (720 BC),' *Biblica* 71 (1990), 206–225.

—, 'The Kingdom of Judah under Josiah,' *TA* 18 (1991), 3–71.

Neugebauer, O., *Mathematische Keilschrifttexte*, I, Berlin 1935.

Neumann, J. and Sigrist, R. M., 'Harvest Dates in Ancient Mesopotamia as Possible Indicators of Climatic Variations,' *Climatic Change* 1 (1978), 239–252.

Nicholls, R. V., 'Old Smyrna: The Iron Age Fortifications and Associated Remains on the City Perimeter,' *BSA* 53–54 (1958–1959), 35–137

Nougayrol, J., 'Textes hépatoscopiques d'époque ancienne conservés au Musée du Louvre, II,' *RA* 40 (1945–1946), 56–97.

—, 'Nouveaux textes sur *ziḫḫu* (II),' *RA* 65 (1971), 67–84.

Oppenheim, [A.] L., 'Zur keilschriftlichen Omenliterartur,' *Orientalia* 5 (1936), 199–228.

Oppenheim, A. L., 'Assyriological Notes to the Old Testament: חרי יונים = *ḫalla summati*,' *JQR* 37 (1946/47), 175–176.

—, '"Siege Documents" from Nippur,' *Iraq* 17 (1955), 69–89.

—, *Ancient Mesopotamia*, Chicago 1977².

—, 'Neo-Assyrian and Neo-Babylonian Empires,' in H. D. Laswell, D. Lerner and H. Speier (eds.), *Propaganda and Communication in World History*, Vol. I: *The Symbolic Instrument in Early Times*, Honolulu 1979, 111–144.

Pardee, D., *Ritual and Cult at Ugarit*, Atlanta 2002.

Parpola, S., 'The Forlorn Scholar,' in F. Rochberg-Halton (ed.), *Language, Literature, and History: Philological and Historical Studies Presented to E. Reiner*, New Haven 1987, 257–278.

Parpola, S. and Watanabe, K., *Neo-Assyrian Treaties and Loyalty-oaths*, Helsinki 1988.

Paterson, A., *Assyrian Sculptures: Palace of Sinacherib*, The Hague 1915.

Pavlov, D. V., *Leningrad 1941: The Blockade*, Chicago-London 1965.

Pickworth, D. 'Excavations at Nineveh: The Halzi Gate,' *Iraq* 67 (2005),

295–316.

Pimouguet-Pedarros, I. 'L'apparition des premiers engins balistiques dans le monde grec et hellénisé: un état de la question,' *REA* 102 (2000), 5–26.

Pinches, T. G., 'Notes upon Some of the Recent Discoveries in the Realm of Assyriology, with Special Reference to the Private Life of the Babylonians,' *JTVI* 26 (1893), 123–171.

—, 'Some Late-Babylonian Texts in the British Museum,' *RT* 19 (1897), 101–112.

Pohl, A., *Neubabylonische Rechtsurkunden aus den Berliner Staatliche Museen*, I, Rome 1933.

Postgate, J. N., *The Governor's Palace Archive*, Hertford 1973.

—, *Fifty Neo-Assyrian Legal Documents*, Warminster 1976.

Powell, M. A., 'Merodach-Baladan at Dur-Jakin: A Note on the Defence of Babylonian Cities,' *JCS* 34 (1982), 59–61.

—, 'Masse und Gewichte,' *RLA* 7 (1987–1990), 457–517.

—, 'Identification and Interpretation of Long Term Price Fluctuations in Babylonia: More on the History of Money in Mesopotamia,' *AOF* 17 (1990), 76–99.

Pritchard, J. B., *The Ancient Near East in Pictures*, Princeton 1969².

Prichard, J. B. (ed.), *Ancient Near Eastern Texts Relating to the Old Testament*, Princeton 1969³.

Rainey, A. F., 'A Front Line Report from Amurru,' *UF* 3 (1971), 131–149.

Reade, J., 'Ideology and Propaganda in Assyrian Art,' in M. T. Larsen (ed.), *Power and Propaganda: A Symposium on Ancient Empires* (=Mesopotamia 7), Copenhagen 1979, 329–343.

Reiner, E., 'New Light on Some Historical Omens,' in K. Bittel et al. (eds.), *Anatolian Studies Presented to H. G. Güterbock on the Occasion of his 65th Birthday*, Istanbul 1974, 257–261.

Risdon, D. L., 'A Study of the Cranial and other Human Remains from Palestine at Tell Duweir (Lachish) by the Wellcome-Marston Archaeological Expedition,' *Biometrika* 31 (1939), 99–166.

Rofé, A., 'The Laws of Warfare in the Book of Deuteronomy: Their Origins, Intent and Positivity,' *JSOT* 32 (1985), 23–44.

de Romilly, J., *Thucydides and Athenian Imperialism*, Oxford 1963.

Rosén, H. B., 'Questions d'interprétation de textes juridiques grecs de la plus ancienne époque,' in J. Modrzejewski and D. Liebs (eds.), *Symposion 1977, Vorträge zur griechischen und hellenistischen Rechtsgeschichte*, Köln-Vienna 1982, 9–32.

Roth, M. T., 'The Material Composition of the Neo-Babylonian Dowry,' *AfO* 36/37 (1989/90), 1–55.

Sachs, A. and Hunger, H., *Astronomical Diaries and Related Texts from Babylonia*, Vol. I: *Diaries from 652 B.C. to 262 B.C.*, Vienna 1988.

Saggs, H. W. F., 'The Nimrud Letters, 1952, Part I: The Ukin-zer Rebellion and Related Texts,' *Iraq* 17 (1955), 21–56.

—, 'Babylon,' in D. Winton Thomas (ed.), *Archaeology and the Old Testament Study*, Oxford 1967, 39–56.

—, *Assyriology and the Study of the Old Testament*, Cardiff 1969.

Salonen, E., *Die Waffen der Alten Mesopotamier*, Helsinki 1965.

Salvini, M., 'Une hypothèse sur le 'Tell des ramparts' de Mari,' *MARI* 5 (1987), 628–630.

el-Sāmarrāie, H. Q., *Agriculture in Iraq During the 3rd Century, A.H.*, Beirut 1972.

San Nicolò, M., *Babylonische Rechtsurkunden des ausgehenden 8. und des 7. Jahrhunderts v. Chr.*, Munich 1951.

Saracino, F., 'A State of Siege: Mi 5 4–5 and an Ugaritic Prayer,' *ZAW* 95 (1983), 263–269.

Sarna, N., 'Zedekiah's Emancipation of Slaves and the Sabbatical Year,' in H. A. Hoffner (ed.), *Orient and Occident (Essays Presented to C. H. Gordon on the Occasion of his 65th Birthday)*, Neukirchen-Vluyn 1973, 143–149.

Sauček, V., 'Ein neues hethitisches Ritual gegen die Pest,' *MIO* 9 (1963), 164–174.

Scheil, V., 'Nouvelles notes d'epigraphie et d'archeologie assyriennes,' *RT* 36 (1914), 179–192.

Schulman, A. R., 'Siege Warfare in Pharaonic Egypt,' *Natural History* 73 no. 3 (March 1964), 12–21.

Scurlock, J. A., 'Assyrian Battering Rams Revisited,' *SAAB* 3 (1989), 129–131.

—, 'The Euphrates Flood and the Ashes of Nineveh (Diod. II 27.1–28.7),'

Historia 39 (1990), 382–384.

Shiff, L., 'Neo-Babylonian "Interest-free" Promisory Notes,' *JCS* 40 (1988), 187–194.

Shiloh, Y., 'The Population of Iron Age Palestine in the Light of a Sample Analysis of Urban Plans, Areas, and Population Density,' *BASOR* 239 (1980), 25–35.

—, 'Underground Water Systems in the Land of Israel in the Iron Age,' in A. Kempinski et al. (eds.). *The Architecture of Ancient Israel from the Prehistoric to the Persian Periods*, Jerusalem 1992, 275–293.

Sigrist, M., 'Seven Emar Tablets, in *kinattûtu ša dârâti* (R. Kutscher Memorial Volume), Tel Aviv 1993, 165–184.

Skaist, A., 'The Chronology of the Legal Texts from Emar,' *ZA* 88 (1998), 45–71.

Slotsky, A. L., *The Burse of Babylon*, Bethesda, MD 1997.

Snodgrass, A. M., 'A Corinthian Helmet from the Persian Siege Ramp at Palaepaphos, in F. G. Maier (ed.), *Alt-Paphos auf Cypern: Ausgrabungen zur Geschichte von Stadt und Heiligtum, 1966–1984*, Mainz am Rhein 1985, 43–49.

von Soden, W., 'Aššuretelilāni, Sînšarriškun, Sînšum(u)līšer und die Ereignisse im Assyrerreich nach 635 v. Chr.,' *ZA* 58 (1967), 241–255.

—, *Bibel und Alter Orient*, Berlin-New York 1985.

Spalinger, A. J., 'A Canaanite Ritual Found in Egyptian Reliefs,' *JSSEA* 8 (1977), 47–60.

Stager, L.E., 'Phoenician Carthage – The Commercial Harbour and the Tophet,' *Qadmoniot* 17 (1984), 39–49 (Hebrew).

—, 'Merneptah, Israel and Sea People: New Light on an Old Relief,' *EI* 18 (1985), 56*-64*.

—, 'When Canaanites and Philistines Ruled Ashkelon,' *Biblical Archaeology Review* 17/2 (1991), 24–43.

Starkey, J. L., 'Excavations at Tell el Duweir, 1933–1934,' *PEF QSt* 1934, 164–175.

Starr, I., *Queries to the Sungod: Divination and Politics in Sargonid Assyria*, Helsinki 1990.

Steiner, R. C. and Nims, C. F., 'Ashurbanipal and Shamash-shum-ukin: A Tale of Two Brothers from the Aramaic Text in Demotic Script,' *RB* 92

(1985), 60–81.

Steinkeller, P., 'Battering Rams and Siege Engines at Ebla,' *NABU* 1987/27.

Stolper, M. W., 'Late Achaemenid Legal Texts from Uruk and Larsa,' *BaM* 21 (1990), 559–622.

Streck, M., *Assurbanipal und die letzten assyrischen Könige bis zum Untergange Niniveh's*, Leipzig 1916.

Stronach, D. 'Notes on the Fall of Nineveh,' in S. Parpola and R. M. Whiting (eds.), *Assyria 1995*, Helsinki 1997, 307–324.

Sukenik (Yadin), Y., '"Engines Invented by Cunning Men",' *BJPES* 13 (1946–1947), 19–24 (Hebrew).

Tadmor, H., 'The Campaigns of Sargon II of Assur: A Chronological-Historical Study,' *JCS* 12 (1958), 22–40, 77–100.

—, 'Philistia Under Assyrian Rule,' *BA* 29 (1966), 86–102.

—,'Rabshakeh,' *Enc. Miqr*. 7 (1976), 323–325.

—, 'History and Ideology in the Assyrian Royal Inscriptions,' in F. M. Fales (ed.), *Assyrian Royal Inscriptions: New Horizons in Literary, Ideological and Historical Analysis*, Rome 1981, 13–33.

—, 'Sennacherib's Campaign to Judah: Historical and Historiographical Considerations,' *Zion* 50 (1985), 65–80 (Hebrew).

—, 'The *urbi* of Hezekiah,' *Beer-sheva* 3 (1988), 171–178 (Hebrew).

—, 'On the Use of Aramaic in the Assyrian Empire: Three Observations on a Relief of Sargon II,' *EI* 20 (1989), 249–252 (Hebrew).

—, *The Inscriptions of Tiglath-pileser III King of Assyria*, Jerusalem 1994.

Thureau-Dangin, F., *Une relation de la huitième campagne de Sargon*, Paris 1912.

—, *Tablettes d'Uruk à l'usage des prêtres du temple d'Anu au temps des Sèleucides*, Paris 1922.

Tsukimoto, A., '"Ein Eselkopf und 1/4 Kab Taubenmist" (II Kön 6, 25b),' *AJBI* 12 (1986), 77–86.

—, 'Sieben spätbronzezeitliche Urkunden aus Syrien,' *ASJ* 10 (1988), 153–189.

—, 'Akkadian Tablets in the Hirayama Collection (I),' *ASJ* 12 (1990), 177–259.

—, 'Akkadian Tablets in the Hirayama Collection (II),' *ASJ* 13 (1991), 275–333.

Tufnell, O., *Lachish* III*: The Iron Age*, London 1953.

Tufnell, O. et al., *Lachish* IV*: The Bronze Age*, London 1958.

Uffenheimer, B., *Early Prophecy in Israel*, Jerusalem 1999.

Ussishkin, D., *The Conquest of Lachish by Sennacherib*,Tel Aviv 1982.

—, 'Excavations at Tel Lachish 1978–1983: Second Preliminary Report,' *TA* 10 (1983), 97–175.

—, 'The Assyrian Attack on Lachish: The Archaeological Evidence from the Southwest Corner of the Site,' *TA* 17 (1990), 53–86.

—, *The Renewed Archaeological Excavations at Lachish (1973–1994)*, Tel Aviv 2004.

van den Hout, Th. P. J., 'Masse und Gewichte bei den Hethitern,' *RLA* 7 (1987-1990), 517–527.

Vargyas, P., *A History of Babylonian Prices in the First Millennium BC,* 1. *Prices of Basic Commodities*, Heidelberg 2001.

de Vaux, R., *Ancient Israel: Its Life and Institutions*, London 1965.

Virolleaud, Ch., *L'astrologie chaldéene,* Paris 1905–1912.

Vita, J.-P., 'Warfare and the Army of Emar,' *AOF* 29 (2002), 113–127.

Walker, C. B. F., 'Cuneiform Tablets in the County Museum and Art Gallery, Truro, Cornwall,' *AfO* 24 (1973), 122–127.

Waschow, H., *4000 Jahre Kampf um die Mauer: Der Festungskrieg der Pioniere (Geschichte der Belagerungstechnik*, 1), Bottrop 1938.

Weidner, E. F., 'Keilschrifttexte nach Kopien von T. G. Pinches: 1. Babylonische Privaturkunden aus den 7. Jahrhundert v. Chr.,' *AfO* 16 (1952–1953), 35–46.

von Weiher, E., *Spätbabylonische Texte aus Uruk,* II, Berlin 1983.

Weinfeld, M., 'The Worship of Molech and of the Queen of Heaven and Its Background,' *UF* 4 (1972), 133–154.

—, 'The Counsel of the "Elders" to Rehoboam and its Implications,' *Maarav* 3 (1982), 27–53.

Welten, P., *Geschichte und Geschichtsdarstellung in den Chronikbüchern*, Neukirchen-Vluyn 1973.

Winckler, H., *Die Keilschrifttexte Sargons nach den Papierabklatschen und Originalen neu herausgegeben*, Berlin 1889.

Winter, I. J., 'Royal Rhetoric and the Development of Historical Narrative in Neo-Assyrian Reliefs,' *Studies in Visual Communication* 7/2 (1981), 2–38.

Wreszinski, W., *Atlas zur altägyptischen Kulturgeschichte*, II, Leipzig 1935.

Xella, P., 'Un testo ugaritico recente (*RS* 24.266, *Verso*, 9–19) e il "sacrificio dei primi nati",' *RSF* 6 (1978), 127–136.

Yadin, Y., 'Hyksos Fortifications and the Battering Ram,' *BASOR* 137 (1955), 23–32.

—, *The Art of Warfare in Biblical Lands in Light of Archaeological Study*, New York 1963.

—, 'Some Aspects of the Strategy of Ahab and David (I Kings 20; II Sam. 11),' *Biblica* 36 (1965), 332–351.

—, 'Weapons, weaponry' *Enc. Miqr.* 5 (1968), 931–970.

Yaron, R., 'Redemption of Persons in the Ancient Near East,' *RIDA*, 3e série, 6 (1959), 155–176.

Yeivin, S., 'The Tunnel in the Lachish Glacis,' *EI* 1 (1951), 29–31 (Hebrew).

Young, T. C., 'The Assyrian Army on the Middle Euphrates: Evidence from Current Excavations,' *Bulletin of the Society for Mesopotamian Studies* (Toronto) 6 (1983), 19–32.

Yurco, F. S., 'Merneptah's Canaanite Campaign,' *JARCE* 23 (1986), 189–215.

Zaccagnini, C., 'An Urartian Royal Inscription in the Report of Sargon's Eighth Campaign,' in F. M. Fales (ed.), *Assyrian Royal Inscriptions: New Horizons in Literary, Ideological, and Historical Analysis*, Rome 1981, 259–295.

—, 'War and Famine at Emar,' *Orientalia* 64 (1995), 92–109.

Zawadzki, S., 'The Economic Crisis in Uruk During the Last Years of Assyrian Rule in the Light of the So-called Nabu-Ušallim Archives,' *Folia Orientalia* 20 (1979), 175–184.

—, *The Fall of Assyria and Median-Babylonian Relations in the Light of the Nabopolassar Chronicle*, Poznań 1988.

Zimhoni, O., *Studies in the Iron Age Pottery of Israel: Typological, Archaeological and Chronological Aspects*, Tel Aviv 1997.

INDEX OF SUBJECTS AND NAMES

INDEX OF SOURCES

Assyrian and Babylonian Sources

www.ingramcontent.com/pod-product-compliance
Lightning Source LLC
LaVergne TN
LVHW010555300926
849162LV00002B/339